The book of PUZZLING Sc

an EDUCATIONAL PUZZLE BOOK

Educational book for Children

Puzzling Science and Maths *and a few other things...*

Publisher's Note

We are delighted to be publishing this book that will interest children and adults alike. Mathematics and science are a source of wonder, fun and enlightenment. Too often they are seen as difficult, boring or worse. Enjoy the book - and see our website for more titles and puzzles to make you wonder!

www.tarquingroup.com

75 puzzles and illustrated throughout

The book of PUZZLING Science and Maths

Published by Tarquin

Suite 74, 17 Holywell Hill

St Albans, AL1 1DT, UK

Book ISBN : 978-1-911093-51-0
EBook ISBN: 978-1-911093-52-7

www.tarquingroup.com

Printed in Poland

Foreword

This is our second inspired book of family based educational puzzles. We found that our puzzles needed a lot more space to offer a full explanation *(not just an easy answer)*. So in this book each puzzle we have devoted two whole pages. My intent was originally to create a series of printed-cards so that children could pick and choose a puzzle, *(so none are in any particular order)*. Chaos for a book you might think, but children often prefer a little chaos before order is re-stored. Our puzzle explanations are easily found, many of which are web linked for additional information *(which I have checked as being suitable for children)*. Un-like other puzzle books, our detailed answers are not hidden away at the back. At the same time the opportunity *is there to* 'THINK' independently about a particular problem is far more important than getting the answer 'right' *(it's not a school test)*. Some of us often think about problems for years.

Not all puzzles have easy *(tick-box)* solutions. This book has taken several years to produce from the initial idea with my son Joshua who also came up with many of the puzzle challenges. Puzzles do have a timeless appeal for every gen-eration and solving them does not make it any less a puzzle for others. Some *'grown-ups'* think everything is 'known' when if fact what we know is a little jigsaw part. One of the recent changes I have made to the book is to change the age guidance (from 8+) as my children are a lot older now and we all have something to learn - so it's just a guide for parents. I make no apologies if some of these puzzles are considered hard to do. Some are easy but others are for when the child gets older. So I hope you'll enjoy this book because of that. In my puzzles children get to learn something that they would not normally think about. It is the more interesting book of Science and Maths and a few other interesting observations thrown in for thought, discussion and challenge!

Philip *and* Joshua Searle

Puzzles re-invented for those that don't usually like puzzles!

Designed to encourage original thought and problem solving.

Index of Puzzles

That's all folks !

Charlie's Extraordinary Glass Elevator RIDE!

With apologies to Quentin Blake

Remember when Willy Wonka took Charlie high above the chocolate factory in his 'Great Glass Elevator' - that then shot straight upwards into the sky and then into 'outer space'?

Did you ever 'WONDER' how fast that lift really went?

As they are not **'floating in space' that** must mean that they are standing in a lift that is going fast enough to keep their feet safely on the ground of the lift - due to acceleration.

That is the puzzle challenge!

Find out the **average speed** (or velocity) of that Elevator - if I tell you that 'Space' is only (about) 60 miles above your head and you don't want Charlie standing in an elevator (going up) forever...

Only turn over when you have: (1) read the book (2) been in an elevator (3) are not afraid of heights (4) are hopelessly Lost in Space!

How fast was that great glass Elevator?

This is similar to **'Rocket Travel'**. To make this simpler (as it's Rocket Science it's not directly comparable to 'real' Rocket launch) but if we take into account the three main key points along the trajectory until we reach 'STOP' (height above the Earth,) we can get an idea of how fast it is. Unlike a rocket, our space 'Elevator' can 'stop' in mid-way, which is a nice idea, if we can ever get a 'Space Elevator' to work! It's not such a daft idea!

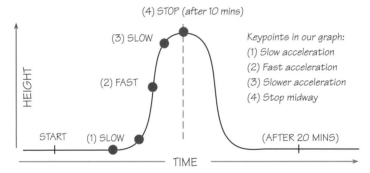

How long would YOU spend in a lift - if you were Charlie?. Let's try 20 minutes total; 10 minutes there and 10 minutes back. Take a look at the graph above - its NOT a proper Graph for a Rocket, (as Rockets don't stop mid-way). It's Charlie's Elevator plotted over time. The main key points are a (1) Slow start (2) Steady acceleration (3) a Slow down and (4) STOP (midway).

We are going to assume that the 'mass' of the Elevator remains constant as we reach our height in 10 minutes flat (halfway). To simplify further I am going to average out the Slow-to-fast segment of the graph and imagine it was a Car or Train on the Ground. If it were a Rocket it would be travelling much faster at the 'top' of the graph than the 'bottom' section! Plus that SPACE actually is just (100km or 60 Miles) away. Rockets need an acceleration of 9.81 $^{-2}$ metres per second per second (or 30 feet/per second per second) to escape the Earth's gravity or they would fall back to Earth!

How fast that great Glass elevator was travelling is very, very fast!

Average speed (in MPH*)= Distance (60 miles) over Time (10 minutes**).
* Miles per Hour. ** !0 minutes is also 1/6 of an hour.
Calculate: **Average speed** = distance 60 (divided by) time (1/6) = 60 x 6 = **360 mph**.

This might be fast for an Elevator but NOT fast enough for a Rocket to get into orbit! But one day a 'Space Elevator' might indeed be possible by 2050.
http://www.space.com/14656-japanese-space-elevator-2050-proposal.html

From these Children's Letter blocks find the missing words - from the words letters you cannot see!

Each block has four letters per cube, one letter per side, (none at top or bottom). There are just four blocks to play with. You should try to make as many words as you can from all the letters you find.

As you can see, there is ONE WORD already - so you need to find the other missing words. And we'll give you a 'clue' - as there are THREE letters missing, so it cannot be that hard can it?

The answer to this puzzle is finding the Words you can see at the FRONT and then add the ones you cannot 'see' on all the other SIDES...

How? The letters you cannot 'see' include 'A and I'. The first word is HERO and with the new letters (shown above) you can easily make another word appear.

Take as long as you like, there is '**no trick**', it's a puzzle and the winner is the first to make four simple words from these four children's letter blocks.

Answer is overleaf - but the puzzle is in the 'doing' here (do not turn over unless you have created at least four words).

Answers to the Children's Letter blocks finding the words you cannot see from the letters you cannot see...

Well we did have some easy clues:
The letters you could not see included **[A] and [I]**

Cube sides:

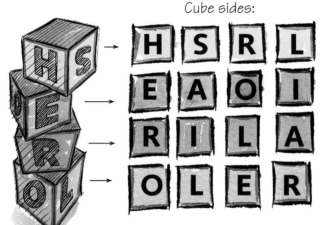

BLOCK FACES:
Left to right.
No letter repeats
on any of the
blocks across.

Note there are NO
LETTERS on any
of the top or
bottom of the
blocks. Only the
sides have
LETTERS printed.

BLOCKS top to bottom: No letter repeats vertically

The First word you see is HERO but using the above letters [A,] [L]
and [I] you should quickly find SAIL. As none of the letters are repeated
on a letter cube you have all the letters you need for more.

Note there are no 'wrong' words - if you have 'other' letter combinations.
Your words may differ from ours - but as long as there is a word on each side
that reads vertically (up or down) we think that is a winning answer.

After all you would not want them to be too easy to do. If you like this puzzle
you can spot something in Newspapers that works the same way (but is probably
easier) as they give you all the letters and you try to make as many words as
you can from four letters and above. Try this version and if you succeed on that
you are ready to move on to other word letter puzzles, (try The Times or Telegraph
junior sections). We reckon you can make 12 words (or more) from the same
letter combinations. **Have fun!**

Other combinations (with OREH) are LATE, LAER, LIAR, LAER, PSIL, LAES, SRAE, SREH, SLIA, LIAH, ELOS, ESOR, LAEH, EROS, RAEL, LAEH, ESIR, etc. (Another beginner any letter in a word). And yes we have scrambled the letters for you. No cheating.

Solve the CRAZY BOX colours!

This is a folding box. The colours inside this box are the same as outside - but what happens when we fold it up?. **Find the TWO boxes that match below!**

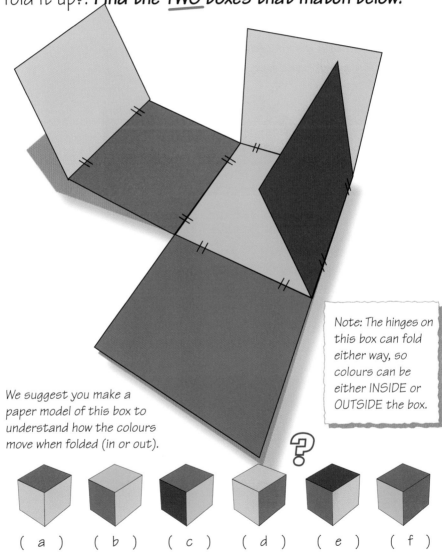

Note: The hinges on this box can fold either way, so colours can be either INSIDE or OUTSIDE the box.

We suggest you make a paper model of this box to understand how the colours move when folded (in or out).

(a)   (b)   (c)   (d)   (e)   (f)

The answer is overleaf. Now try another puzzle box similar to this one...

Authors comment: This was an odd idea from a 'dice' where colours replace numbers.

The answer is (a) and (c). Now try and solve this empty cardboard (upside down) box which has just fallen on the floor... A puzzle perhaps?

What colour do you think is 'hidden' on the other side of this cardboard box...

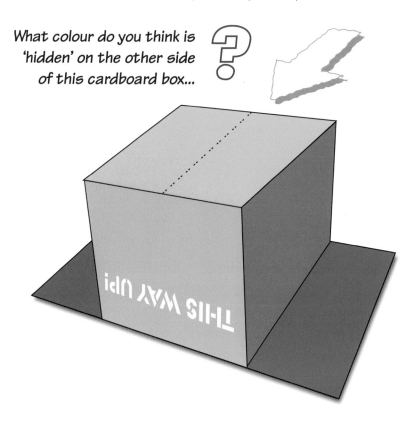

Is it one of these FIVE box colours?

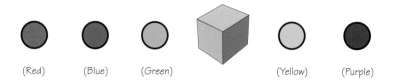

(Red)   (Blue)   (Green)   (Yellow)   (Purple)

Answer (see page 27 to see how to decode this answer):
Llew ti smees dnik fo suoivbo taht revetahw ruoloc si etisoppo thgil eulb sah ot eb thgir. Neve fi ti's edispu nwod dna edisni tuo! Refer ot eht suoiverp egap ot dnif tuo tahw eht etisoppo thgil eulb si. Dnoces xob elzzup devlos.

THE *tricky* CAKE PUZZLE

The three cakes represent three circles. All you have to do is divide equally the three cakes between four boys!

A = 5" dia.

B = 4" dia.

C = 3" dia.

The three cakes are in 5:4:3 ratio. You need to show how these can be equally divided amongst four hungry boys. The cakes are all equal thickness, so they differ in width only (not height). If you're careful you should be able to do this in three slices. So each of the four boys get exactly the same amount each.

There is a clue I can give you. Each boy gets his share in two pieces. By making some good guesses you should be able to work out how to do it to avoid a scuffle of 'who ate the most cake' and 'who had the least'. Some possible solutions are:

(1) Yes you could **'mash-them-all-up'** and weigh them! (but could you do that in 3 slices?)

(2) You could also work out the **volume** of each circle (from the bun width shown 5 inch, 4 inch and 3 inch*).

* As we are all 'Metric', I could make these cakes in Kilometres - but for simplicity 1 inch = 25.4mm

(3) Using LOGIC (and geometry) that Henry Ernest Dudeney used in 1917. He found a simple solution to the problem using ratios.

But how do you do that? (Answers are overleaf)

THE *tricky* CAKE PUZZLE

There is a simple solution.. As each of the cake's width has an exact ratio of 3:4:5 we can draw this triangle:

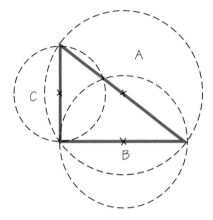

The three circles equal three cakes and as they only differ in width (not height) we can draw a right angled triangle based on these measurements in a ratio of 5, 4 and 3.

It follows (using Pythagoras*) that the size of the two smaller cakes (C and B) is exactly equal to the larger cake A.

So that's a start. Let's name the boys instead of (A, B, C, D) as (A) Alan, (B) Bertie, (C) Charlie and (D) Dick. As they are all very hungry - we had better hurry up and solve this puzzle quick...

FIRSTLY we give Alan and Bertie the two halves from cake A (circle A) - so they will have half of each of the largest cake.

NEXT for the other two boys (C) Charlie and (D) Dick get the remaining cakes split into two halves. (Circles B and C). Each half being equal, they get one half from each slice.

This works because of the ratio of each cake to the other is clearly defined as being 5:4:3.

Circle A

Circle B

Circle C

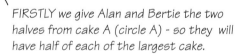

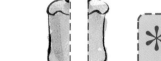

* The proof is via Pythagoras based on simple ancient geometry Maths called the 'Lune of Hippocrates'; you can find out more below:

Credit: This puzzle is attributed to the work of Henry Ernest Dudeney (Amusements in Mathematics - 1917)

http://www.cut-the-knot.org/Curriculum/Geometry/PythagoreanLune.shtml

Getting connected!

The aim of this puzzle is to draw in all the spokes to each hub. Lines must not cross each other, and all hubs must connect to each other in one network. Hub No. 8 spokes have already been drawn in for you to get you started...

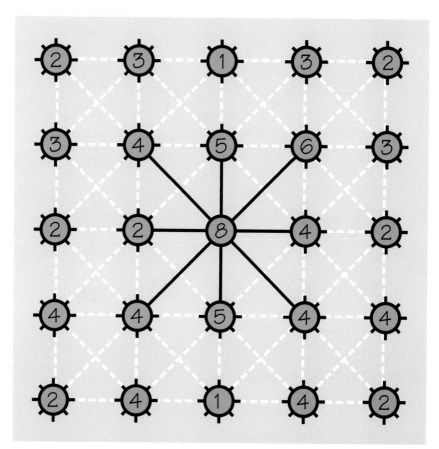

Based on a puzzle by Dave Tuller

You can use a soft Pencil to connect up the Hub. Remember to only connect the correct number of spokes to the hub i.e. No. 8 has eight spokes and number 2 has 2 spokes etc.

Solution?

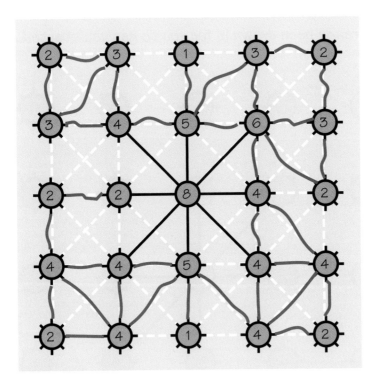

There may be more than one way to do this (we think), so you won't be wrong if yours looks a bit different!

As you can see it's not that easy!
(Imagine if you could NOT cross a gas, electricity or water pipe in your entire street!

PROPORTIONS OF THE PLANET EARTH TO THE MOON USING ANCIENT GEOMETRY!

START with a right angled triangle!

(1) Draw a right angled triangle with the proportions of 3, 4 and 5

You will need a sheet of paper, pencil, ruler and a compass

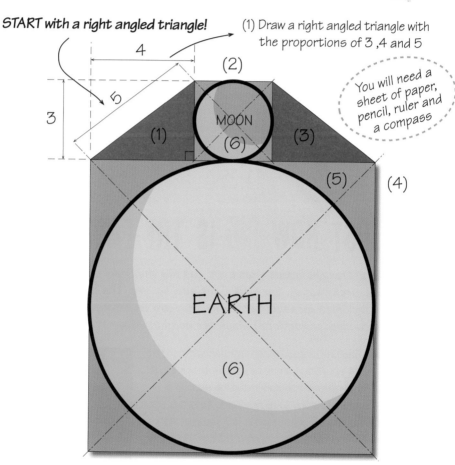

(2) Draw a square box matching the height of the triangle

(3) Draw an identical triangle to tne first one 'flipped' over

(4) Draw a base line from triangle to triangle and make a square below

(5) Draw a diagonal line from corner to corner both sides

(6) Draw a circle where the lines cross, repeat for small square above

So what is the RATIO of Moon to Earth, and how BIG do you think the Moon is anyway? Turn over for more puzzling measurements...

THE DIAMETER OF THE EARTH TO THE MOON

The ancient Egypyians used Cubits as a measurement for observation. Satellite Navigation is now measured in Kilometres which would indicate that the rato was marginally 'off' by 0.1%. Near accurate observation by ancient astronomers!

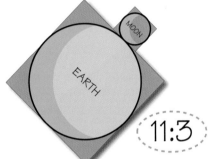

(a) Using the same units divide the smaller into the larger should give you the RATIO 11:3

(b) Using this RATIO you can now find how BIG the Moon and how BIG the Earth is!

AND JUST HOW BIG IS THE EARTH?

The Egyptians thought about that a lot, (as did the Greeks and the ancient Stonehenge architects). So we'll continue to use the original ancient Egyptian/Babylonian measurements (as they are an interesting example of ancient astronomy projections) which are good approximations to this day.

The Egyptians had calculated by careful observation that the sun takes 24 hours to complete a circle of the Earth (they did not know the Earth was rotating around the Sun). Even so, a handy thing to remember is that it takes 86,400,000 Cubits. Using the ratio 11:3 you can work out the size of the Moo'ns circumference from the Earth measurements. You have both ratio and measurement.

Credit: picture: Creative Commons

YOUR ANSWER SHOULD BE IN CUBITS!

To convert Cubits to UK International Nautical Miles* for Navigation on the Earth (or on the Moon) by remembering that 1 degree of arc (Which on any 360 degree compass is one degree) = 240,000 Cubits = 1 Nautical mile for all ships at sea, a familar number. One degree of latitude is approximately 60 nautical miles. This has been 'standardised' as being 1,852 metres (or 6,076 feet). In terms of land miles this is approximately 1.16 statute miles. All UK measurements are now European (French) metric SI units since 1971.

National Physics Laboratory: a history of Measurement:
http://www.npl.co.uk/educate-explore/posters/history-of-length-measurement

Fridge Magnets

This is an easy puzzle once you know how it's done. There are no 'trick solutions', you just need to know where to place the pieces...

Problem: move just TWO pieces to make this point upwards (instead of downwards).

To help you; try to find some counters, sweets, pennies or round fridge magnets to try out a few possibilities.

Solution over page. But you may find another way, be imaginative!

Fridge Magnets

Don't panic! It's easy, FIRST just take the bottom (No. 1 Red counter) and place it on the top line between Yellow and Blue counters.

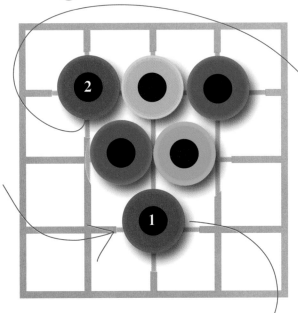

Note: you can use any colour piece. It does not have to be Red!

SECOND: Move the other (No.2 Red Counter) below the top Blue counter (next to the Green). There you have done it!

Or did you cheat?

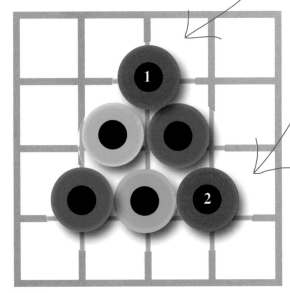

Leap frog logic game

The Green Frogs are swopping places wiith the Blue Frogs. But they can only move ONE SPACE at a time.

One frog can leap over another to land on an empty space and all other frogs can slide forward or back.

You can do this by cutting out the frogs (frog counters left) and placing them on an imaginary board to work out how they could move.

Note: Don't worry about the size of the squares (or the size of the board) it is how they move that is important in this puzzle. It will need quite a few moves on either side to finally get the Frogs to the other side of the board.

Answers shown overleaf - if you're stuck!

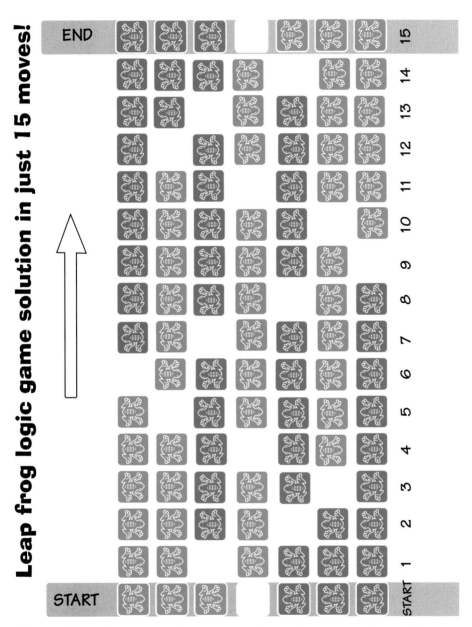

If you like this puzzle game there is an online video version which explains how it's done!

http://www.youtube.com/watch?v=MRXdxnsogxg

How 'HEAVY' is a Rain Cloud?

Every minute of the day on average roughly 960 million tonnes of rain falls on the Earth!

Is it not puzzling how all that weight got into the air in the first place... ?

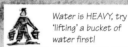

Water is HEAVY, try 'lifting' a bucket of water first!

The thing about water is that it can simply *evaporate* as water vapour from the heat of the sun warming the oceans -(or the air)- creates tiny water droplets of vapour. Every cloud in the sky is water vapour and (as clouds float by) are held by the surrounding air in suspension.

Clouds can be vast, or tiny puffs, each floating or 'blown' by the wind (Clouds do not 'make the winds' although when wind blows clouds form very easily when air pressure changes). That happens a lot in England - you see Clouds and if the conditions are just 'right' that condenses into a downpour of rain.

But even that may not be enough! - Microscopic specks of sea salt, dust or even tiny **bacteria** are needed to 'seed' the growth of a tiny cloud droplet as it condenses from water vapour. Then the cloud droplets have to crash and collide with each other to grow into a raindrop — **about a million cloud droplets are needed to create a typical raindrop**. The raindrop has to be large enough to overcome the forces keeping it up in the air.

The slightest change of air pressure of a cloud being 'pushed' over a mountain top or valley range is enough to form rain. As you know the higher you go 'up' in the air the colder it gets, and the warm air being cooled (by being pushed uphill) may be enough to make vapour condense into rain on the side of a mountain. This often makes mountain 'valleys' flow with water.

So you know how HEAVY a Cloud is. Now what about a raindrop? (See overleaf)

How 'HEAVY' is Rain?
RAIN is heavy but what about a small raindrop?

Raindrops start off as round shapes and grow larger turning into a 'hamburger' shapes, with a flattened base and a round top, as they push against the air — this causes 'drag'. Larger raindrops can become so distorted that can eventually tear themselves apart into two smaller drops. We can see only the 'splatter' of a raindrop stuck on a glass window - which slowly becomes the familiar 'teardrop' shape we recognise*.

 Sphere

 Hamburger

 Air-pressure distorts

 Splits into two

*A shower of rain often unleashes a variety of raindrops, though. This was discovered in 1904 by an American farmer, Wilson Bentley, who laid out a pan of flour to collect falling raindrops — when the rain hit the flour it formed pellets of dough that could be measured for size.

The 'drag' of the air slows the speed down (also prevents raindrops from killing anyone). If a large raindrop fell only 500m (1,640ft) through a vacuum it would smash into the ground at nearly km/h (225mph), and in a rain shower that would be disastrous. The air's 'drag' slows the rain, and then a large raindrop (the size of a small fly) — about 5mm (0.2 in) across (as a sphere)— falls at roughly km/h (20mph). So you're quite safe. But then when the wind blows rain in your face, it may seem to be travelling faster! And of course clouds don't form in a vaccum.

▌PUZZLE TIME: Can you now find the number of raindrops you'll need to fill a bucket?

Well, now you know about rain. How many raindrops do you think are in a bucket of water? Weigh the bucket and capacity then divide the volumes of a raindrop into it. If a raindrop is 5mm (0.2 inch approx) across (make a sphere*) For simplicity raindrops are simply 'spherical' for this puzzle.

You will need to know both the weight and the volume of water in that bucket (it may be written on the side of the bucket (i.e. 1 gallon or 5 Litres), if not you will have to measure that (use a measuring jug) and weigh-it (subtracting the weight of the bucket afterwards). Logically you realise that the weight of the water (in the bucket) must be in proportion to the volume of water (as raindrops). It will be tiny - but you have now solved the puzzle. Raindrops are as water, relatively heavy. Congratulations you've solved the puzzle!

*Weblink: http://studymaths.co.uk/formulae/volumeofasphere.html

Leonardo's script!

Did you know...
Leonardo da Vinci put some of his work in mirror writing so no-one could read it? But it is possible for you to do the same and even make it a bit harder!

Each of these few lines contain a hidden message and they will get harder and harder to read. – this is the standard text.

(1) Ha ha. This text is the upside down, I've fooled you – or have I...?

(2) This is impossible to read...unless you have a mirror which means you are very clever. hmm.

(3) This is upside down and in mirror writing. This is far too difficult for you. Or is it?

(4) This sentence looks easy but it's too hard for you to crack!

(5) Yet was one last bit that the last one was rather difficult for you; but this one you will too fail you for ever. All the words are kept to from left as well as upside down and deforming!

How many did you work out?
```
0 = Try harder
1 = Novice codebreaker
2 = Average codebreaker
3 = Good codebreaker
4 = Excellent codebreaker
5 = Master of all codebreakers
```

Perhaps you can find the deliberate spelling mistakes we made as well?.
Just how many ways are there to 'read' a message? See overleaf for more hidden codes.

Be a Codebreaker!

your chance to be a proper codebreaker...

Throughout history there have been clever ways to write a secret message. The Romans it appears swapped secret messages to their legions by wrapping a ribbon around a pole and writing on it. The ribbon then could be carried swiftly by courier (well as fast as a horse could run) and then handed over to a 'staff' member. Only a member of 'staff' could read the message as they had an identical 'staff pole' that they could rewrap the ribbon around and display the message to the General. Ingenious Eh?

But there are so many more codes to choose from and ways to hide them. For a start our alphabet can be changed to another Alphabet entirely. The American commanders in the last war are tributed with an uncrackable code that partly used the language of Navahoe indians (as nobody outside America would recognise the language).

And then there are codes hidden in the symbols of ancient Egypt. All have meaning if you can read it. Although practically few can these days, it was never a hidden code to begin with.

And then we have the plight of the British government secret spy code lost down the back of a fireplace. Only now one ex military historian found the answer and once explained it was obvious (or was it?).

AOAKN HVPKD FNFJW YIDDC

RQXSR DJHFP GOVFN MIAPX

PABUZ WYYNP CMPNW HJRZH

NLXKG MEMKK ONOIB AKEEQ

WAOTA RBQRH DJOFM TPZEH

LKXGH RGGHT JRZCQ FNKTQ

KLDTS FQIRW AOAKN 27 1525/6

IS IT SOLVED?
A Canadian World War II enthusiast says that he has deciphered the message after realizing that a code book held the key to the encryption. Gord Young, (editor for the history group Lakefield Heritage Research), said the 1944 note uses a simple World War I code to give information about German troop positions in the area around Normandy, France.

http://www.abroadintheyard.com/was-carrier-pigeon-discovered-in-surrey-fireplace-bringing-home-secret-coded-message-from-normandy-1944/

class exercise (on computer):

Although you would never know it (looking at your computer), you can also write backwards, upside down, reverse and more as you type. It can get a bit confusing so that is probably why they left it off (but you clever reader can access the hiidden codes and make your own SECRET MESSAGE by typing your message into this website and converting all the characters inside out or upside down (or mirror) just as Leonardo da Vinci did! (Try it yourself on link below).

weblink: http://textmechanic.com/Reverse-Text-Generator.html

Lewis Carroll logic squares

Lewis Carroll is better known for his 'Alice in Wonderland' book but he was also a Mathematician and created some unique Logic puzzles like this one...

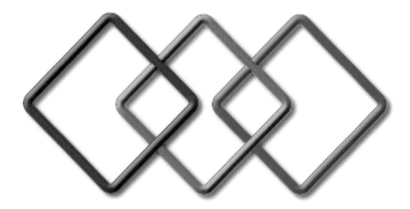

All you have to do to solve this puzzle is to trace each square in one continuous loop linking all the squares together in one go and not tracing any part more than once.

Slightly more complicated (but as easy when you know how it's done) is this four square to solve in the same way.

Some solutions overleaf...

Lewis Carroll **logic squares**

A solution (and there is usually more than one way to do this). You will <u>not</u> be wrong if yours is different to ours.

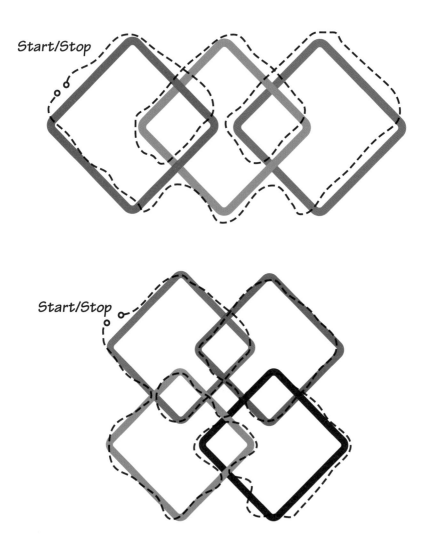

Start/Stop

Start/Stop

If you would like to see more Lewis Carroll puzzles (not all are as easy as this one) find out more. Click on the weblink below.
http://www.knowl.demon.co.uk/page36.html

$\mathcal{D}udeney's$ The Pentagon and the Square

'A regular pentagon may be cut into as few as six pieces that
will fit together without any turning over and form a square,
as I shall show below. Hitherto the best answer has been in
seven pieces, the resolution produced some years ago by a
foreign mathematician, Paul Busschop.' However in 1912 Henry
Ernest Dudeney was able to demonstrate this puzzle...

As you can see Dudney
'improved' on the idea using
just SIX PIECES, and it fits
perfectly as it should.

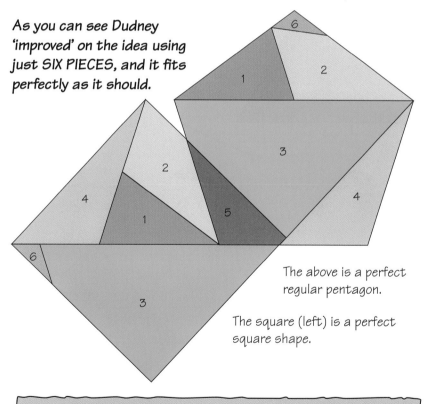

The above is a perfect
regular pentagon.

The square (left) is a perfect
square shape.

If we take any similar sided figures with equal area then we can dissect
these into a finite number of pieces to form each other. This is called the
Wallace-Bolyai-Gerwien Theorem (in case your ever asked) !

NEXT page we make a triangle to a square, we can recreate the 'Haberdashers problem'.
When you can see two rectilinear figures (with equal area) they can both be dissected
into any number of similar pieces to form each other again. See example overleaf...

Dudeney's | **The Pentagon and the Square**

**Now we will see if we can convert a triangle to a square.
You can do this by cutting out this triangle paper shape
and re-arrange the pieces as shown below.**

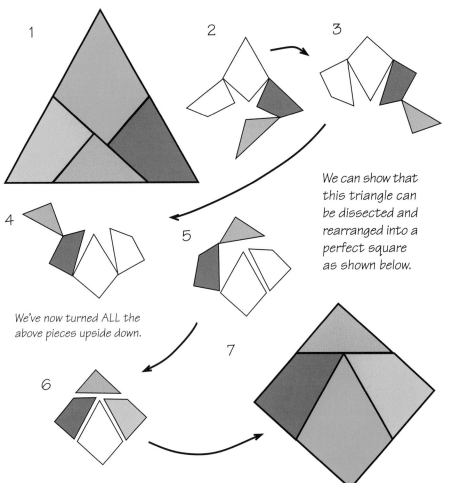

1

2

3

4

5

We can show that
this triangle can
be dissected and
rearranged into a
perfect square
as shown below.

We've now turned ALL the
above pieces upside down.

6

7

Credit: This puzzle is attributed to the work of Henry Ernest Dudeney (Amusements in Mathematics - 1917)

There you are - isn't that simply amazing?

**There are more shape shifting puzzles like this waiting to be discovered
and this is one of our favourites: http://www.cutoutfoldup.com/index.php**

INCREDIBLE
WATER...

A glass of WATER is a perfect science puzzle!

What **EXACTLY** is it?
'H_2O'

There are at least EIGHT* things in a glass of water that have unusual properties that make life on earth possible...

If you can discover just two 'odd' SIMPLE observations **in a glass of water** then you are on your way to discovering a lot more - just by looking at it. DISCOVER more unique properties of water and write them down before you turn over the page and see more puzzling observations you may have missed...

* *There are so many interesting things about water we could not fit any more on the page overleaf!*

INCREDIBLE WATER
The answer is incredibly clear!

SURFACE TENSION
1. The surface tension of water is much higher than liquids of a comparable density, because of strong force between the Hydrogen and Oxygen atoms (H_2O). You can float a paper clip on the surface of a glass of water or an insect, such as the Pond Skater can move about easily on the surface with small feet.

PERFECT FUEL
2. Water is a perfect fuel. Explosive Hydrogen and Oxygen are used to launch NASA space rockets into orbit. So you could say it is a water rocket If it carried water (which it doesn't); it has to have liquid Hydrogen and liquid Oxygen (both frozen with liquid Nitrogen) to keep them as a gas until ignited, the 'by product' at the hot end is just water vapour!

WATER IS HEAVY
3. Water is very heavy which is surprising when Hydrogen and Oxygen are so light. Hydrogen gas is 16 times lighter than Oxygen gas. Hydrogen balloons would float up into space and disappear on their own. Airships did use Hydrogen but now use Helium instead which is very expensive because it has to be distilled. Helium is a rare gas in our atmosphere and not as light as Hydrogen.

FROZEN WATER
4. Frozen water can form beautiful Snow structures of delicate ice crystals; each one unique which form at precisely the right temperature (about 4 degrees) in the humidity of the clouds up above. Too cold and we may get sleet and too warm we may get rain. Despite water being heavy, one large cloud can hold thousands of tons of water. i.e. One inch of rain over one square mile equals 17.4 million gallons of water (weighing nearly 72,000 tons). And yet the clouds gracefully float in air and are blown by the wind!

WATER EXPANDS
5. Water can freeze quickly or slowly at 0 degrees - it does vary in freeze tempreature according to local Air pressure and water quality. For example, pure water can be made to freeze well below zero. If it has one tiny shock (or knock) it can freeze almost instantly.

WATER BOILING POINT
6. When trying to boil water for a cup of Tea on a very high mountain (not that I have personally tried this yet) the Air pressure will make the water boil at a far lower temperature than 100 degrees C. The water will boil when warm and with a lower air pressure it just boils away into vapour and becomes another cloud. So when Water freezes (at 0 degrees C) or boils (at 100 degrees C) it is not totally accurate but a guide only.

WATER SOLVENT
7. Water dissolves and combines to create a solution more than any other liquid, one of these is the mineral Sea Salt of our oceans. When salt is added to water it makes it less dense and lighter, fresh rain (river and glacial) fresh water is heavier and helps create strong ocean currents such as the 'North Atlantic Conveyor' which helps make the UK warmer than other Northerly countries at the same latitude.

WATER and LIFE Sciences
8. Water enables the Photosynthesis of plants to contribute oxygen to our atmosphere by absorbing CO^2 and locking the Carbon into plant growth (such as Trees). Water is the 'enabler' for transporting Nutrients around the plant by dissolving important Minerals from the roots to the leaves. Almost all plants are 'solar powered' this way. Some plants i.e. Algae are important Oxygen producers and simply float on oceans and (in turn) are eaten by tiny Plankton (an important food source for Whales for example). It's all part of a primary food chain.

Water is incredibly puzzling isn't it? There is so much to learn from such a 'simple' substance that behaves in a very complex way. There are many questions you may ask from simple observation in a glass of water. Why does it cling to the sides of a glass? and why does it seem to bend light? is a good observation to explore. Ask questions, find answers!

More things about water: http://www.lsbu.ac.uk/water/anmlies.html

Durer's Magic squares...

Albecht Durer was a German Artist and he cut this picture into a woodcut
(an early form of printing); he died in 1528. The picture has hidden
properties and a sequence of numbers that can be calculated to 34...

16	3	2	13
5	10	11	8
9	6	7	12
4	15	14	1

The picture shows many odd perspectives but one of the most intriguing
is the apparemt random numbers which all add up the same in any direction.
Vertical, Horizontal and diagonally as well as in any group of four numbers
and in any direction (it seems) they all add up to 34.

This adds to the mystery of the picture. There are more hidden number
combinations that have been discovered since. Also hidden in this picture
is the date of its creation of 1514. The sequence of numbers remain
unique to this day. See if you can find more variations before turning over.

http://www.artcyclopedia.com/artists/durer_albrecht.html

Did you find all Durer's Magic squares?

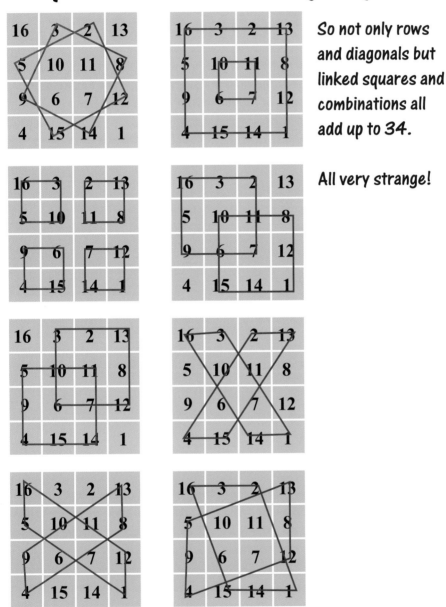

So not only rows and diagonals but linked squares and combinations all add up to **34**.

All very strange!

There are many more than I have shown here.
How many variations? - **Nobody really knows!**

'Mi Marvellus bruvver'

...has removed all the four tin-can labels and stuck on his own labels (all mixed up of course). But with a little bit of LOGIC by opening ONLY ONE CAN we can find out which is which...

(Not Apricots)   (Not Dog Food)   (Not Soup)   (Not Biscuits)

So that is the puzzle - to find out what was in each tin (before label swap) and which to open first when all the labels are mixed up seems an impossible challenge!

Vital clues:

(1) All tins are the same size. (2) Each tin has the 'wrong' label stuck on. (3) Imagine these are 'real' tins- you can pick them up and inspect them. *So what would you do?*

You must solve this puzzle before your Mum finds out and 'explodes' (you will get the blame as usual for your brother or sister).
See if you can solve it before your Mum does overleaf!

Based on a book by Ian Livingstone and Jamie Thompson (How Big is your Brain?).

LoGIC Stops Mum from 'exploding'

This is all you need to do to stop your Mum from exploding.... and it's easy!

First - by picking up the biscuits -are they the lightest?
If not they cannot be full of any biscuits!

So put aside that tin and check the contents of one of the other tins by opening just one tin. By checking contents (i.e. label SOUP and inside it's DOG FOOD) this is a vital clue because that means that the other two cans left must contain the SOUP or APRICOTS. But which one?

Finally you realise that of the two cans left one must contain (say) SOUP, the other must contain (say) APRICOTS so if you simply swop the labels of the two, that should solve the problem.

Now you know what is in each tin-can and you can change the labels back to what they should be -before your Mum finds out and explodes!

VICTORIAN ARITHMETIC AND IMPOSSIBLE SUMS!

$$45 - 45 = 45$$

An impossible sum! - (you may think) Yet this Victorian curiosity was taught in Maths lessons to baffle school children...

First we make an observation that forty-five is made up from all numbers one-to-nine like this: 1+2+3+4+5+6+7+8+9 (which equals forty-five). Secondly: reverse the figures (as they make no difference to the total) and take one from the other.

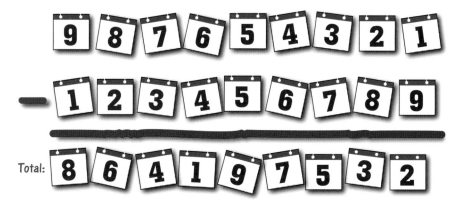

9 8 7 6 5 4 3 2 1

− 1 2 3 4 5 6 7 8 9

Total: 8 6 4 1 9 7 5 3 2

By subtracting the smaller from the larger number (which are still 1 to 9 exactly as before) and then adding them all up the total is still 45. Perhaps your Maths teacher can explain why this is so?

Most Amazing: 6174

Choose any *FOUR* digit number (make sure the digits are *NOT* all the same!)

EXAMPLE: 9 1 8 9

- REARRANGE TO HIGHEST NUMBER: 9 9 8 1

- REARRANGE TO LOWEST NUMBER: 1 8 9 9

- SUBTRACT LOWEST FROM HIGHEST: (8 0 8 2)

- REPEAT ONCE MORE AND AGAIN...

UNTIL YOU FIND, (YOU GUESSED IT).   6 1 7 4

Wow!

Amazing because this works with ANY four digit number!

Credit: Thanks to Kjartan Poskitt for allowing us to show this amazing number puzzle! www.MurderousMaths.co.uk

How Eratosthenes used the Sun to Measure the Earth!

Eratosthenes was a Librarian working in Alexandria (in 205 BC), he was able to demonstrate that you could use the 'Summer Solstice' to calculate the circumference of the World!

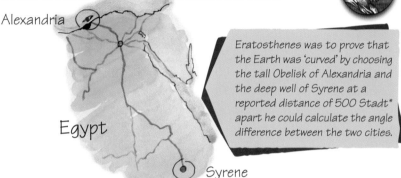

Alexandria

Egypt

Syrene

Eratosthenes was to prove that the Earth was 'curved' by choosing the tall Obelisk of Alexandria and the deep well of Syrene at a reported distance of 500 Stadt* apart he could calculate the angle difference between the two cities.

The plan was to see if one site had a shadow (when the Sun is directly overhead) when the other site did not.? This caused some confusion, if the Earth were indeed flat, why a shadow at all?. On the other hand if the Earth was round, (as Pythagorus and Archimedes had long suspected); as a ship at sea could see 'more', the higher up the mast you go was well known, but nobody had proved it was true.

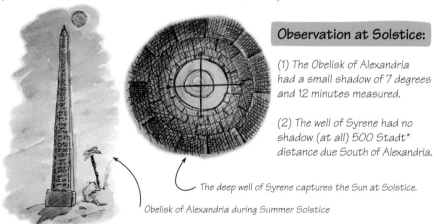

Observation at Solstice:

(1) The Obelisk of Alexandria had a small shadow of 7 degrees and 12 minutes measured.

(2) The well of Syrene had no shadow (at all) 500 Stadt* distance due South of Alexandria.

The deep well of Syrene captures the Sun at Solstice.

Obelisk of Alexandria during Summer Solstice

* 500 Stadt = (500 Roman Stadium pitches) = 279Km in total. Eratosthenes employed a 'pacer' to walk the entire distance from Alexandria to Syrene. It was found to be surprisingly accurate despite the winding road and Syrene not being at the Equator or even in-line vertically with Alexandria.

So how do you measure the Circumference of the Earth? See overleaf!

How YOU can calculate the circumference of the world!

The formula for calculating the circumference of circle is 2 x (pi) x radius. But we have to find another way (as we do not know the radius yet). Eratosthenes only made his calculation after Thales of Militus* had already calculated the height of a Pyramid (from a distance without actually measuring it directly).

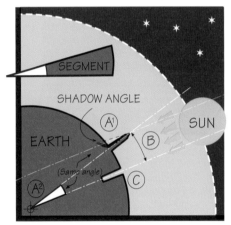

By using the difference in the length of the shadows (during Summer Solstice), Eratosthenes could calculate the difference in the angle at which the sunlight hit the northern most point (in Alexandria), and the sunlight that hit the southernmost point (in Syrene).

This angle of difference is a tiny section of a 360 degree circle. The distance between the two sites (500 Stadt) is a tiny slice of the earth's total surface.

(A²) The shadow angle (7 degrees 12 minutes) is from the tip of the Obelisk of Alexander from the vertical projection made back down to the Earth, as shown above.

(B) 500 Stadt is the exact distance between Alexandria and Syrene.

(C) The well of Syrene is shadowless during the Summer Solstice.

The Sun rays fall equally on both sites at the same time. To calculate the Earth angle we divide the 7 degrees 12 minutes into 360 degrees of a perfect circle (which is 1 / 50th of the World.) Just multiply that number of segments by 500 and that will give you the circumference of the World.

So what is the circumference of the EARTH in Stadts?

An astronomical observation made 3,000 earlier (BC) had enabled Stonehenge and the Pyramids of Egypt to be accurately aligned to the stars and to predict the seasons as 'natural measurement' for Earth time (based on the circumference of the Earth). Eratosthenes much later in 205 BC discovered himself the circumference of the Earth that had long been 'forgotten' along with the Pyramid architects. It was however well known by the ancient Egyptians who built the Pyramids and the English Druid architects of Stonehenge (as Stonehenge was completed 500 years before the ancient Pyramids of Egypt were constructed).

*Thales of Militus is better known for his 'Thales Theorem' on angles in a Triangle:

The circumference of the world was calculated by Eratosthenes in 205 BC
http://www.the-map-as-history.com/demos/tome10/11-circonference_demo.php

What happens when the **ICE** melts?

This is a tumbler full of water with ice-cubes added.

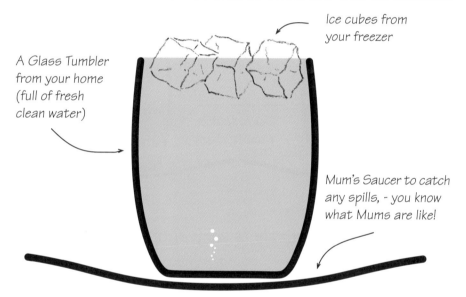

Ice cubes from your freezer

A Glass Tumbler from your home (full of fresh clean water)

Mum's Saucer to catch any spills, - you know what Mums are like!

The puzzle is very simple...

Choose from one of these options:

(1) The ICE raises the water level in tumbler

(2) The ICE reduces the water level in tumbler

(3) The water level remains the same

ANSWER (coded). See if you can decode the answer. (See page 27)

Eht rewsna si taht ti sniamer ta eht emas level. Sgrebeci gnitaolf ni eht aes od ton knis tub taolf esuaceb eht retaw (sa eci) sdnapxe. Nehw eht eci stlem ti ylpmis sekat pu eht ecaps ti dah decalpsid ni eht aes. Os on egnahc ot retaw level.

What happens when we MIX Water with ALCOHOL?

This is a half tumbler full of water + a half tumbler of Alcohol. If we mix one with the other of equal volume, what do you think happens?

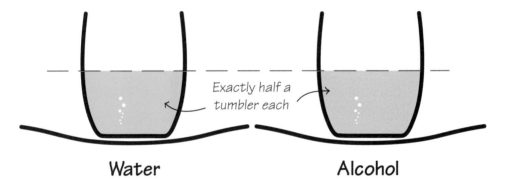

Exactly half a tumbler each

Water   Alcohol

The puzzle is very simple...

Note: Alcohol is inflammable, *dissolves stains on your carpet and you could run a car on Alcohol instead of Petrol. (This has nothing to do with the puzzle here, just thought you should know 'health and safety' inspectors would not allow you to do this experiment at home or at school). It can only be done in a Laboratory where they can measure 'volumes' precisely. Not your mum's kitchen scales!*

Choose from one of these options; **(when MIXED)** *what happens?*

(1) The ALCOHOL raises the water level in tumbler

(2) The ALCOHOL reduces the water level in tumbler

(3) The MIXED levels remains the same

ANSWER (coded). See if you can decode the answer. (See page 27)
Lohocla sah a laiceps ytreporp taht sdnob htiw retaw ta a ralucelom level taht sekat pu ssel ecaps naht retaw flesti. Rof elpmaxe si ew ekat eno ertil fo retaw dna eno ertil fo Lohocla dna ew xim meht neht eht emulov fo retaw si ssel (ta 1.629 sertil).

PYRAMIDS AND CUBITS IN SPACE
(OR MEASUREMENTS IN TIME AND SPACE)

The Egyptian Cubit is shown to be more accurate than the Metre and has been for over 4,000* years. Not only is the Cubit accurately related to the true size of the Earth (The metre failed that test) the Cubit is also related to a mathematical constant called (pi).

Gold!

The Cubit still exists (ironically in France - the home of the Metre) at the 'Place de la Concorde' called the **'Obelisk of Luxor'**. It was a gift from the Viceroy of Egypt in 1831 and was made at the same time as the Great Pyramids of Egypt, weighing 227 tons made of solid red tint granite and is over 3,300 years old. It is exactly 76 feet tall - (equivalent to 23.1648 metres).

(which is exactly 50 Cubits*)

One Cubit was reputed to be the length of an average mans arm. Modern reference books have now defined this as 18.24 inches precisely. So we can state that the **Obelisk of Luxor** is 50 Cubits that converts to 12" (old standard inches) to a foot x 76 = 912. If we then divide 912 by 18.24 inches then we get 50 Cubits exactly - the height it was made .

50 Cubits

The Obelisk of Luxor

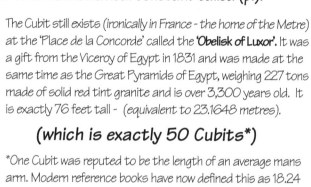

Pyramids were designed and made using Cubits. The great Pyramid of Egypt called 'Cheops' is 756 feet at the base (which converts approx to 500 Cubits*).

Pyramid of Cheops

500 Cubits

*approximately as subject to sand-blast erosion.

Observant Londoners may know of Cleopatra's Needle (can be seen on the Thames Embankment in London) which weighs 187 tons. *The Cubit is also the oldest measurement on record by the Egyptians (2800- 2300BC).

The word obelisk is Greek meaning 'object' and the top of the obelisk was usually gold (missing from our version in London which is covered with Pigeon poop!)

Are you ready for LIFT OFF yet? It's even more puzzling overleaf...

PYRAMIDS AND CUBITS IN SPACE

WHEN THE FRENCH MEASURED THE EARTH NORTH POLE TO EQUATOR THEY THOUGHT THEY HAD ESTABLISHED A 'NATURAL' CIRCUMFERENCE...

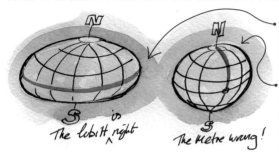

The Cubit right

The Metre wrong !

• Ships' Navigators know now that the Equatorial circumference of the Earth is 21,600 Nautical miles.

• The French however decided in 1792 to measure from the Equator to the North pole and then made the metre one ten-millionth of that measurement.

Puzzle credit: based on historical data by Nigel Corrigen at the www.dozenalsociety.org.uk

We know that the Earth is not perfectly spherical (fatter at the Equator) so the new base measurement for the Metre has been made three times since then becoming more odd...

In 1960 the French Material 'standardised' the Metre as a wavelength of 1,650,673.72 of the 'reddish-orange' light emission from an isotope of Krypton-86Krb gas and then again in 1983 how far light can travel in vacuum at 1/299,792,458 of a second. Ghastly!

This is a very long way from an 'EARTH' inspired natural measurement (the original intention of the French Metre). The ancient Egyptians had earlier defined the Earth as 86,400,000 Cubits (which is a number we recognise of some importance for Navigators at sea). This number represents exactly (in thousands) how many seconds in a day (24 hour day).

You need to know exactly where you are at sea. For this reason you always Navigate with a 360 degree compass. Each degree has 60 minutes, and each minute 60 seconds. These are useful measurements of time and space for Ships, Aircraft and Spaceships.

The Egyptians knew that the Planet was 86,400,000 Cubits (around the equator), then 1 degree of arc (on your compass) = 240,000 Cubits and 1 minute equates to one international nautical mile. This is exactly the same Nautical mile used today.

0/360°   *Ships Compass*
270°   N
W   E   90°
S
180°

The Sun moves in Cubits, the Earth receives 24 hours sunshine (if it were seen overhead in a spacecraft from space) that would be 86.4 million Cubits in total. In one hour the Sun's shadow moves at speed of 3,600,000 Cubits (one minute 60,000 Cubits and in one second, 1,000 Cubits). Astonishingly accurate measurement of the speed of the Earths rotation!

24 hours = 86,400,000 Cubits

Ah 360°!

Even more interesting is that the Cubit is a natural constant measurement in Maths...
If we see our world as having 360 degrees of a circle then dividing (18.24 inches x 2) = $\sqrt{9.868}$ which gives us the square root of (pi) as **3.141337294** etc (or 22/7 approximatly). This is because 'the Obelisk of Luxor is 50 Cubits that converts to 12" (old standard inches) to a foot x 76 = 912. If we then divide 912 by 18.24 inches then we get 50 Cubits exactly.'

THIS PUZZLE WAS SOLVED BY THE ANCIENT EGYPTIANS!

Weblink: http://www.dozenalsociety.org.uk/history/inch.html

WHY does my computer have an odd keyboard?

About 'QWERTY' Keyboards...

It is no coincidence that your computer keyboard has QWERTY KEYS (and almost every keyboard you have ever seen). You might think it faster if it had NORMAL keys with A, B, C, D, E, etc. - and you would be right!

The QWERTY typewriter KEYBOARD was invented prior to 1878*, and a number of inventors sought to develop a reliable keyboard that did not suffer from 'clash-of-keys', (when two or more keys get locked together trying to spell out a word). Other keyboards existed for Music writing (notation) or for the Blind (Braille) but the early office typewriter was still a mechanical device that involved each letter being 'hit' with a miniature hammer with a single letter on each to 'stamp' it's way onto a sheet of paper. Typing fast 'was a problem' as it would jam. The solution was to re-arrange (mix up) the keys so that fast typists would 'slow down' their typing speeds to suit the machine speed!

Early Typewriters ...

From 'non' QWERTY keyboard (left) to QWERTY (right) which has become the 'standard' keyboard layout but there were plenty of 'others' invented as you can see oveleaf.

*Find out more about typewriter keyboards inventions overleaf...

WHY are my computer keys all jumbled?

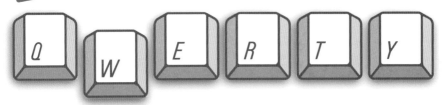

So how are you getting on with your own keyboard? Can you really type fast? Well, there are 'other'alternatives which have never quite 'caught on' but offer a unique experience over the QWERTY keyboard. What do you think of these below?

More strange Typewriter keyboards!

We would like to show you how these worked but it is much better to leave it to your imagination!

By default almost all 'TYPEWRITERS' could not change the letter style (such as italic or bold) until IBM invented the 'GOLFBALL' typewriter which offered better print quality and change of styles (such as italics). This was swiftly followed by the 'DAISYWHEEL' typewriter. At the same time both 'GOLFBALL'and DAISYWHEEL' had built in 'data storage' using either punched paper tape or (much later) a computer floppy disk for letter (memory) storage.

Golfball (letters stuck on a ball shape)

And finally when computers became 'affordable' they replaced the standard typewriter model with your (non impact) computer printer. The last UK Typewriter factory (Brother Limited located in North Wales) was closed as recentlly as 2009 - but the QWERTY keyboard lives on, but for how long is anybody's guess!

All images curtesey of Creative Commons

Daisywheel (letters stuck on a spinning wheel)

Explore the history of the typewriter on these weblinks below:
http://www.typewritermuseum.org/history/why.html
http://home.earthlink.net/~dcrehr/whyqwert.html

DRAW or PHOTOCOPY twelve L shapes and make a large rectangle simply by re-arranging it. After 10 minutes and you can see no solution then you're as bad as Herbert...

Oh you're SO clever Clogs (thinks Herbert).

All at sea - **Ship Clock chimes**

Ships' clocks are different; they chime on the hour (only up to four times per hour) and on each half hour.

So our puzzle question is this...

How many chimes are there in total on a Ship's clock between the time of 12.30 day and 12.30 at night. Remember that ships' clocks 'chime' once on the half hour and twice (double bell) on the hour.

You can easily work out what time it on board ship if you already know your 'am' or 'pm' (Latin for midday - 'ante meridiem' - and after midday - 'post meridiem').

12:00 8 bells (four double bells)
1:00 2 bells (one double bell)
2:00 4 bells (two double bells)
3:00 3 bells (three double bells)
4:00 8 bells (four double bells)
Plus a single bell is chimed on the half hour (every hour).

Ships' clocks need to be more accurate primarily for navigation. 'Four hours on and four hours off' working is a normal work shift for those on board ship.
Check your answers overleaf...

All at sea - **Ship Clock chimes**

Answers:

Start:
12:00 MIDNIGHT

12.30 1 bell = 1 bell (1 chime)
1:00 2 bells (one double bell) = (2 chimes)
1.30 1 bell = 1 bell (1 chime)
2:00 4 bells (two double bells) = (8 chimes)
2.30 1 bell = 1 bell (1 chime)
3:00 3 bells (three double bells) = (6 chimes)
3.30 1 bell = 1 bell (1 chime)
4:00 8 bells (four double bells) = (8 chimes)
4.30 1 bell = 1 bell (1 chime)

5:00 2 bells (one double bell) = (2 chimes)
5:30 1 bell = 1 bell (1 chime)
6:00 2 bells (two double bells) = (4 chimes)
6:30 1 bell = 1 bell (1 chime)
7:00 2 bells (three double bells) = (6 chimes)
7:30 1 bell = 1 bell (1 chime)
8:00 2 bells (four double bells) = (8 chimes)
8:30 1 bell = 1 bell (1 chime)

9:00 2 bells (one double bells) = (2 chimes)
9:30 1 bell = 1 bell (1 chime)
10:00 2 bells (two double bells) = (4 chimes)
10:30 1 bell = 1 bell (1 chime)
11:00 2 bells (three double bells) = (6 chimes)
11:30 1 bell = 1 bell (1 chime)
12:00 2 bells (four double bells) = (8 chimes)

12:00 MIDDAY
12:30 1 bell = 1 bell (1 chime)

End:

The number of double bells + single bells over this period is twice that (i.e. 12 hours night time and 12 hours day time). This makes a grand total of 77 bells during the day time and another 77 at night. Total 154 chimes!

Weblink: http://shipsclock.sourceforge.net/intro

THE SCISSOR PROBLEM...

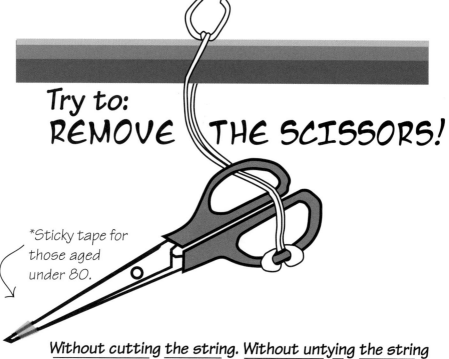

Try to:
REMOVE THE SCISSORS!

*Sticky tape for those aged under 80.

Without cutting the string. Without untying the string -or trying to slide the string off the metal pole!

YES - IT IS POSSIBLE...!

Scissors are sharp of course (but you knew that already), be very careful and we used *sticky tape to close the Scissor ends (as shown School 'health & safety'). We also found a Coat hanger to tie the Scissors to (works better than a metal bar) although you could also use a nearby radiator pipe if you were to challenge somebody else to solve the puzzle!

Have fun!

THE SCISSOR ANSWER!

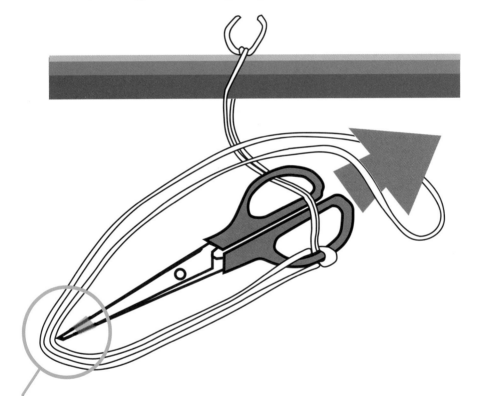

Carefully slide the string over the top of the BLADES LIKE THIS!

Safety POINT is we first tape up the ends of the BLADES. SCISSORS ARE SHARP!

Upside down Words

When you hold up a clear lemonade bottle before your eyes you will see some words appear upside down whilst others do not...

CARBON DIOXIDE

CARBON DIOXIDE

Tip: Hold the lemonade bottle, half way between the paper and your nose!

TOXIC WASTE

TOXIC WASTE

Tip: You could also use a large Magnifying glass, it works the same way!

Just to make it a little harder on the second bottle we have changed one letter. Some words are easier than others.
See if you can make some 'upside down' word combinations of your own called Ambigrams* (Shown overleaf)!

Try some based on Ambigrams*

Here are some good examples and how to make your own from one or two words...

Start with your name or your favourite TV show... I copied from memory 'Vision On' (BBC childrens Art programme) and this was one of Tony Hart's favorite creations which bounced around our TV screens in the 1970's.

I drew this on tracing paper and then traced it off and turned one sheet over and traced both as one. This is really easy to do but you may need to play with your letters first. Some letters work better than others!

*These are just two examples that read the same up or down. Artists such as John Langdon or Erich Friedman are able to create words that read up or down with amazing clarity. These are called Ambigrams.

John Langdon: http://www.johnlangdon.net/
Eric Friedman: http://www2.stetson.edu/~efriedma/ambigram/

Weight lost in Space?

This is an updated version of a Lewis Carroll classic puzzle and it asks an intriguing question...

What *do you* think happens when the Astronaut tries to climb this rope ladder?

Same weight and both in orbit

Both the Astronaut and the Satellite are of exactly the same weight - that is when they both were on Earth!

And for this puzzle piece we are ignoring 'mass of the objects' and Earth's gravity entirely.

Its a question of balance!

- See overleaf for another puzzle...

A question of balance...

*In the original 'monkey' problem outlined by
Lewis Carroll (the Alice in Wonderland Author
and Mathematician) he proposed this problem...*

Well this puzzle
has been argued
over for some
time. If the weight
equals the monkey
then it is perfectly
balanced - but
only the monkey
can climb and the
weight cannot!

**What do-you-think happens when the monkey tries to
climb the ladder? And then what about the Spaceman
(overpage) you saw earlier - what happens there?**

Issac Newton stated that any object of equal mass will balance
in equlilibrium. So in Space if the Astronaut tries to climb the
ladder he will find: (1) the ladder is useless (2) it acts like a rope
and both will be pulled to the centre of the rope.

Credit: (Sam Loyd's Cyclopedia of Puzzles) http://www.jwstelly.org/CyclopediaOfPuzzles/PuzzlePage.php?puzzleid=P244.1

The Centre of a Triangle?
(is surprisingly difficult)

You will need a school compass, a right angle set-square and a pencil to try this ancient Greek puzzle out...

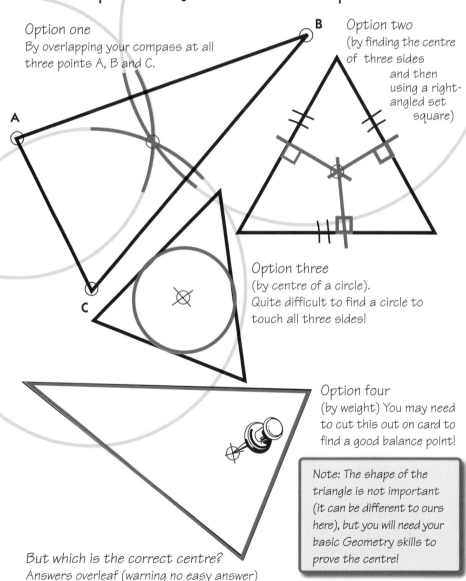

Option one
By overlapping your compass at all three points A, B and C.

Option two
(by finding the centre of three sides and then using a right-angled set square)

Option three
(by centre of a circle).
Quite difficult to find a circle to touch all three sides!

Option four
(by weight) You may need to cut this out on card to find a good balance point!

Note: The shape of the triangle is not important (it can be different to ours here), but you will need your basic Geometry skills to prove the centre!

But which is the correct centre?
Answers overleaf (warning no easy answer)

The Centre of a Triangle?

There are two types of Triangles - Equilateral and all others.
There are hundreds of possible 'centres' depending on how
you define a 'centre'. It's not that easy!

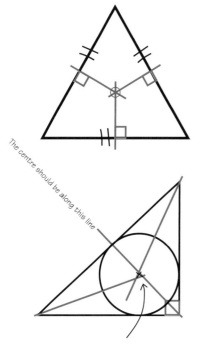

The centre should be along this line

*Lets look at an equilateral Triangle and all
sides being equal simply find the middle of a
line and use a right angle set square to draw
a line. This would be the same even if you used
a compass.*

*However if you think this works for every other
triangle think again...*

*This is a right angle Triangle and if you find the
centre of the angle (45 degree) with a compass
you will get the 'obvious' centre. But then if you
use your same compass and try to make a
circle in the triangle then you will find that the
centre has moved.*

*In fact the centre gets more annoying by moving
whenever the triangle angles are changed. But
that's not all...*

But the centre of the circle is slightly off centre...

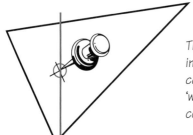

*The more distorted the Triangle the more
impossible it seems to be to find the 'true'
centre. If you again try to define a centre by
'weight' then it is nowhere near any other
compass or circle centre. It is all very puzzling.*

Discover yet more Triangle center problems on this short video link:
http://archive.org/details/journey_to_the_center_of_a_triangle

Dudeney's Triangles to 'SQUARE' 1

Henry Ernest Dudeney was an English mathematician interested in constructing challenging Maths puzzles and this is one of his many geometric puzzles from 1917.

This puzzle is based on rectangles split in half equally. Each rectangle width is exactly half the height. You would need to Photocopy and then cut-out carefully the shapes below and rearrange these resulting triangles into a perfect square...

Be careful with the Scissors!

ONLY CUT ALONG THE DOTTED LINES

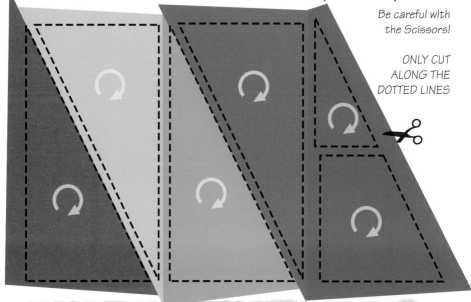

It may be helpful to know that this is fairly easy to solve in that you only need to rotate the six Triangles into the correct shape and not 'flip them' i.e. you do not need to turn them over at all.

We shall be featuring more of Dudeney's geomectric puzzles - but not all are as easy as this one! The answer to this puzzle is overleaf but only if you have tried for 5 minutes or more...

Dudeney's Triangles to 'SQUARE'1

To solve this puzzle is easy! All you need to do is rearrange the Triangles (in any order) of colour in this particular order. There may be other solutions, or even other shapes based on this.

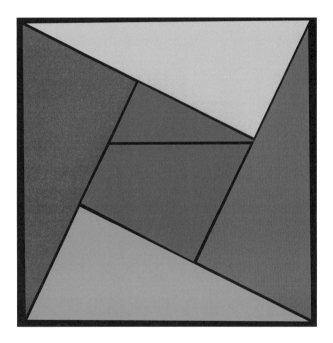

Credit: This puzzle is attributed to the work of Henry Ernest Dudeney (Amusements in Mathematics - 1917)

Note how the (Blue) Triangle is split into two distinct parts to make it fit neatly. I think you'll agree that it is a very neat solution to the puzzle. **And a fun way to explore Maths?**

Both Lewis Carroll and Henry Dudeney had fun playing with odd shapes and creating puzzles and solutions from the world around them. We shall be looking at more Dudeney and Carroll puzzles in our other pages soon.

Can **Triangles** be Patented?

How about a square or a circle or Hexagon?

One man called James Warren (in 1848) made a patent application for a new kind of 'bridge' one not based on old 'square boxes' but instead based on the power of ▲ (as you can see below).

The Warren girder bridge

It was constructed two years later at London Bridge Station. It proved that equilateral Triangles are even better and stronger than similar 'isosceles' triangles. Many such Iron bridges have been built since (particularly railway bridges). Perhaps you can you spot a 'Warren girder bridge' built near you. Both box and Warren girder bridges are a common bridge construction.

See if you can recognise 'other' shapes used in modern construction elsewhere - such as the 'Hexagon' - used on the EDEN PROJECT building (in Cornwall) and elsewhere. Is this a puzzle, **I think so!**

See how this construction has influenced modern architecture overleaf...

Maybe! Designs can patented for all sorts of structures!

Not only Triangles - but there are other shapes used in construction being used. Such as Hexagon and this is a shape favoured by Architects as it can be 'curved' as a dome shape and is strong. Puzzling why it's not used elsewhere - it's an ideal construction method!

(a)

(b)

(c)

The Gherkin in London (top)

British Museum (right)

Eden project (both below)

(a) London's Gherkin tower is basically a Triangular construction (b) inside detail. (c) The British Museum dome roof is clearly built from Triangles. (d) The Devon 'Eden Project' is built totally from Hexagons.

(d)

The row-boat problem

Two girls and two boys are stranded on a desert island with a very leaky boat. The boat will sink if they all get in it, but it will just about float with TWO girls or just ONE boy only.

Plus: They cannot swim to the nearby island due to hungry Sharks nearby. How do they all get across to the other island?

Boys: 7 Stones weight

hungry shark

Leaky boat (sinks with anything over 12 Stones weight)

Girls: both 5 Stones

The two boys are 7 stone each but the girls are just 6 stone each. The leaky boat will sink with anything heavier than 12 stone on board. So only one boy, boy plus girl or two girls can cross. It's up to you how they get across...

It can be done! But it may mean a lot of rowing... The boys and girls cannot swim but both are good at rowing.

Your challenge is to find a convincing way of crossing in the boat so that they can survive without getting wet or being eaten alive by the Sharks! This interesting puzzle was part of Eleanor Searle's School homework (then aged 7).

Solve the (very tricky) island crossing with this desperate solution...

The correct answer to the tricky island crossing:

First the 2 girls go to the other side and 1 gets off and the other rows back to pick up a boy. She rows back to the other side and drops him off. She then rows back and collects the other boy and she gets out. The boy then rows to the other side and he gets out whilst the other girl rows back and picks up her friend - and then (finally) they both row back to the safe island. **Remember this if you get stuck on a remote Island surrounded by Sharks on a leaky boat...**

Who cares about Pythagoras?

You should! If you take any size sheet of paper and fold it diagonally (you get a square). Divide the side length size into the diagonal size and you will find the 'Pythagoras Constant'

Length

Diagonal

It may be called the 'Pythagoras Constant',
but it was **Hippasus** who discovered i
i.e. it is the square root of 2 $(\sqrt{2})$
which is 1.41 *

George Lichtenberg in 1786 had the idea of using this ratio for making different size paper. i.e. making paper in the same ratio would be an efficient use of paper. If we take an A4 sheet for example: (297mm x 210mm), divide 210mm into 297mm = the 'Pythagoras Constant' consistant for all other paper sizes, (see overleaf for 'other' sizes).

A4

* Well the Square root of 2 is a BIT longer than just 1.41...
1.4142135623 7309504880 1688724209 6980785696 7187537694
8073176679 7379907324 7846210703 8850387534 3276415727
3501384623 0912297024 9248360558 5073721264 4121497099... etc.
- this is the truncated version as there are x number of decimal places!

OK. So it has a practical use, where's the puzzle?

ISO (International) PAPER SIZES are all based on the square root of two, or approximately 1:1.4142. This ratio was conceived by Georg Lichtenberg in 1786. At the beginning of the 20th century, Dr Walter Porstmann turned Lichtenberg's idea into a proper system of different paper sizes. And this is because it makes all paper 'scalable' in size with minumum paper waste. It's a bit of a puzzle that this information cost Hippasus his life, and Pythagoras got the credit.

Read why (for Hippasus) it was a dangerous ratio: http://nrich.maths.org/2671

Pythagoras cares for paper!

Based on this Mathematical constant; by using the ISO international paper size we can work out the most economical paper to use for a given size.

The most common print paper sizes are either A or B. The letter designates the series and then is followed by a figure, forexample: A1 is half the size of A0. A0 is the the basic sheet size and A1 is half the longest length. Similarly A2 is half A1, A3 is half A2, A4 is half A3 and so on. The C range is mostly uses in envelope sizes. Standard A (or B) sizes are the most economical print sizes. Printers use 'B' stock paper size and then trim down to 'A' sizes for final delivery.

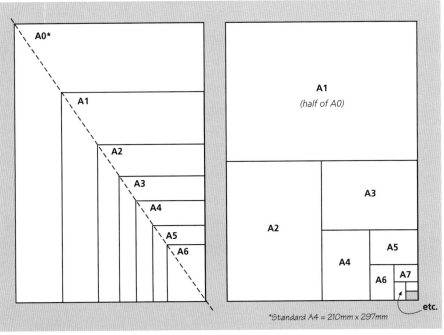

*Standard A4 = 210mm x 297mm

So **Pythagoras** really does care about saving trees and paper by using the most economical paper size by using pure Maths. As you look around you you will find a number of printed items as 'standard sizes' and one of those that are 'not' strictly this size is BOOKS. That is because they are based on old 'imperial' paper sizes or because the publisher wants a book that is another shape to attract your attention. Printers rely on this information to work out how much each book costs.

Pascal's triangle Maths puzzle

It's a bit more PUZZLING than you think!

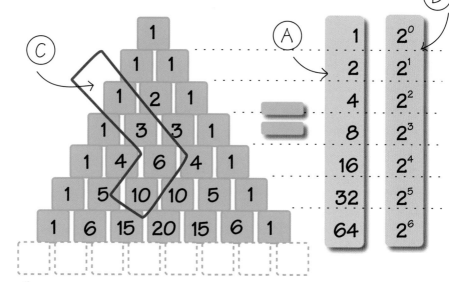

This is the famous Pascal's triangle invented (or discovered) by Blaise Pascal in 1653 (and some say based on an ancient Chinese puzzle dated 700 years earlier). It has lots of 'hidden' functions that are still amazing Maths puzzlers today. The pyramid of numbers does not stop but continues in the same way to infinity.

So the first thing to do is fill in the blanks before we show you puzzles A, B or C - just so you can see how they work...

This simple sequence of numbers are all logical and display how numbers can create some odd calculations that all add up.

(A) = The **doubling sequence** of numbers on each line
(B) = The **'squaring'** (powers of two) of each line
(C) = The hockey stick **reveals the sum** of the three numbers above

But there is so much more to Pascal's triangle than you think...
If you would like to know more then just turn over page...

Pascal's triangle Maths puzzle

Yet more PUZZLING sequences of numbers!

$$1 \quad = \quad 11^0 \quad = \quad 1$$

$$11 \quad = \quad 11^1 \quad = \quad 11$$

$$121 \quad = \quad 11^2 \quad = \quad 11 \times 11$$

$$1331 \quad = \quad 11^3 \quad = \quad 11 \times 11 \times 11$$

$$14641 \quad = \quad 11^4 \quad = \quad 11 \times 11 \times 11 \times 11$$

> Who would have thought the 11 times table is in the Pascalls Triangle?

Discover 'Triangular numbers'

Yes just add the dots in a sequence (as below) and Pascall will add them all up for you!

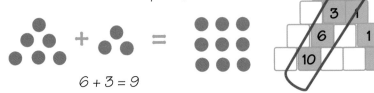

$$6 + 3 = 9$$

Works with consecutive triangular numbers such as $(n + n + 1)$

Discover 'Fibonacci numbers'

These occur in nature (petals of a flower for example) are fibonnaci number sequences by adding up the levels (1) then (1-1) then (1+2) then (1+3+1) then (1+4+3) to find the next number.

$1 + 1 = 2$, $1 + 2 = 3$, $2 + 3 = 5$, $3 + 5 = 8$, $5 + 8 = 13$, $8 + 13 = 21$ *(and so on)*

Discover more about Pascal's triangle on this weblink.

http://www.mathsisfun.com/pascals-triangle.html

ODD MAGIC SQUARE

This square seems to defy any logical explanation.
By re-arranging this square into another the square
hole in the middle disappears!

**Photocopy or print out this page and
cut out carefully all the shapes above.**

Now you can re-arrange the shapes by rotating
90 degrees (twice) clockwise to make a new
large square as shown here and overleaf.

ODD MAGIC SQUARE

The finished square should look like this after rotation...

By twisting each piece twice 90° is the answer. Now what was the question?

You may be able to guess why the central square hole almost vanishes and impress your friends with your discovery!

ELEANOR'S School NUMBER PUZZLE 1

Ask a friend to do this EASY calculation in their head...

1. Think of a number (any number 1 to 10)...

2. Double that number in your head...

3. Add four to that number...

4. Multiply by three...

5. Divide by six...

6. Subtract two from the total

Now what was that number* you thought of?

* It is always the number you started with.

OK that was EASY. We said it would be!

But there is more... **See OVERLEAF!**

ELEANOR'S School
NUMBER PUZZLE 2

Ask your friend to find a BOOK with at least one hundred pages (any paperback book would do that). They will also need a pencil and paper.

1. Tell them to flip to any page and ask them to write that number down.
2. Tell them to choose a line (from one to ten) from that same page
3. Tell them to write that line number down
4. Tell them to choose one word from that line (one of the first nine words).
5. Tell them to write that exact word and the word number down
6. Tell them to take that page number and double it
7. Tell them to multiply that number by ten and add twenty
8. Tell them to add the line number and add five
9. Tell them to multiply that total by ten
10. Tell them to add the the word number

 NOW THEY HAVE A TOTAL - ASK IF YOU CAN HAVE THAT TOTAL. YOU WILL NOW WORK OUT WHAT EXACTLY THAT WORD IS!

To solve this TOP SECRET puzzle you must do the following;
11. Ask for the same book and subtract two hundred and fifty from the TOTAL number they give you. (DO THIS SUBTRACTION SECRETLY)

13. The **FIRST TWO** number are the page number - (or the FIRST number - if you only have a three-digit number.)
14. The **SECOND** number is the line number
15. The **THIRD** number is the word number

Now you can tell them the exact WORD they chose from the book!

A Carpet fitters 'problem'

It's a bit of carpet (remnant) and is an odd shape. But it has to fit a SQUARE bedroom. This is sadly the last piece of carpet on the planet.

Can you find a solution to this tricky problem?

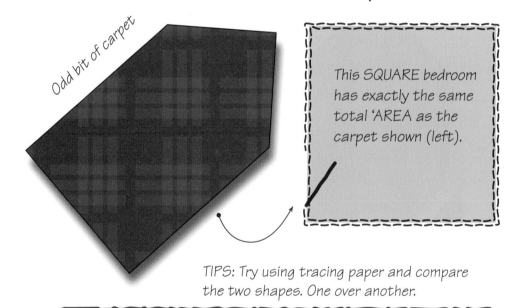

Odd bit of carpet

This SQUARE bedroom has exactly the same total 'AREA as the carpet shown (left).

TIPS: Try using tracing paper and compare the two shapes. One over another.

THINK GEOMETRY! It looks an impossible problem to solve at first. But it can be done with just TWO CUTS to fit that square room exactly - without any wasted carpet!

As normal we show our answer overleaf, (and yet there is more than one way to do it). All are valid if it 'fits' neatly for the customer, (your Mum or Dad perhaps!)

That is the PUZZLE. The Carpet 'problem' to solve...

A Carpet fitters 'problem'

SOLVED (we hope)?

So how did you do? Remember there is (usually) more than one way to solve a problem. We think this is the neatest solution - if they have the same AREA.

The Carpet has to fit somehow. We first straightened up the shape (left) and drew in a centre line and added the two parallel dotted lines...

Then cut-out the bottom of the Carpet into two equal equilateral triangles. (Geometry assumes they would be the same size as the top two triangles).

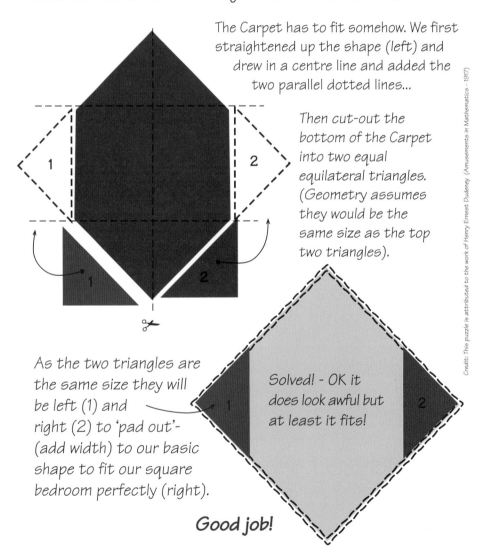

Credit: This puzzle is attributed to the work of Henry Ernest Dudeney (Amusements in Mathematics - 1917)

As the two triangles are the same size they will be left (1) and right (2) to 'pad out'- (add width) to our basic shape to fit our square bedroom perfectly (right).

Solved! - OK it does look awful but at least it fits!

Good job!

Rodney's LOST insurance cost!

Rodney goes shopping to buy a new **Mobile phone** - he has exactly **£60** to spend. He then buys a **Mobile phone** which has 'Insurance' and then buys a **waterproof protective case** to protect his phone from future shark attack. And it's all included in the price...

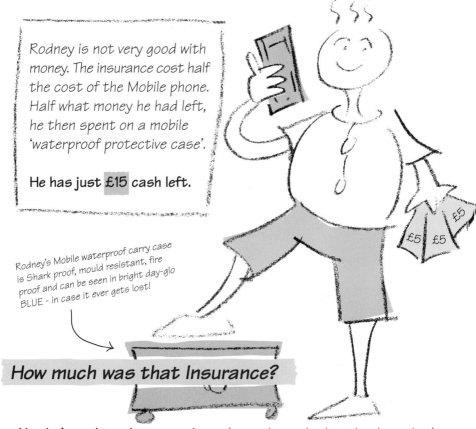

Rodney is not very good with money. The insurance cost half the cost of the Mobile phone. Half what money he had left, he then spent on a mobile 'waterproof protective case'.

He has just £15 cash left.

Rodney's Mobile waterproof carry case is Shark proof, mould resistant, fire proof and can be seen in bright day-glo BLUE - in case it ever gets lost!

How much was that Insurance?

You don't need to ask any questions - *(or need to go back to the phone shop).* It's a small problem and with a little bit of clever Maths and you could find out the cost of the insurance deal.

Our answers are always on the back page of the next page. (No cheating).

Rodney' discovers the true cost...

(1) **Rodney has got £60.**
We also know he has NOT spent 'all-of-it' as he still has £15 left.
So our first calculation is easy (£60 - £15 = £45). He has spent £45.

(2) We also know that he spent TWICE as much on his Mobile phone as his mobile
insurance. (Because his insurance was half the cost of the Mobile phone).
We still don't know the cost of his mobile yet.

(3) As we already know (above) that Rodney has spent TWICE as much on his
Mobile phone as his insurance. Therefore: if he spent £45 in total (which
must include £15 he spent on the carry case). *Quick calculation: £45 overall
expenditure minus £15 on that single purchase = £30 remaining.*

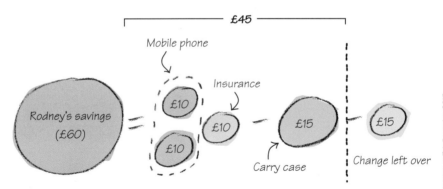

(4) **Finally using a ratio of 3:1 of £30 remaining we find the insurance.**
Split £30 into three and you have two parts Mobile Phone and one part
Insurance. *Because he spent twice as much on his mobile phone as his
insurance; (example: £10 + £10 + £10 = £30).*

So Rodney's Insurance was £10.
Insurance was perhaps expensive. But then his mobile phone had been saved
from fierce Shark attack!

Rodney thinks it was good value. **What do you think?**

Credits: Adapted from the DAILY MAIL 'Mindbenders' puzzle section on 7th December 2015.

The disappearing SQUARE puzzle

Watch the grey square change its colour as you move it...

[a]

This puzzle demonstrates how our eyes (and brain) adapt to subtle changes of light and shade naturally...

(1) Photocopy this page twice (2) Cut-out the square [a] neatly (see square top left). (3) Place a plastic cup over the shadow area (over the circle marked in Red) (4) Take the 'cut-out' square [a] and place over square [b] as shown. (5) Slide the [a] square from [b] to the centre [c] square.... **Do you see the colour change now?**

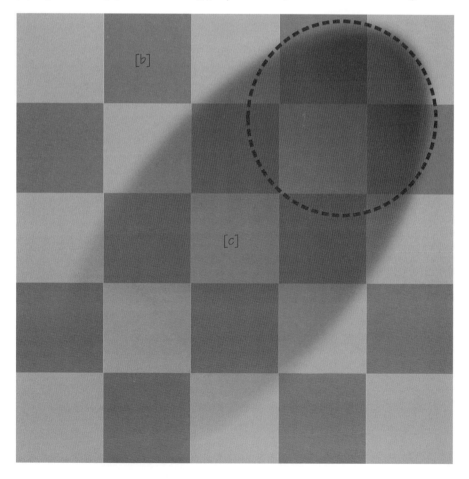

[b]

[c]

PROOF that it changes colour!

Edward H. Adelson

The colour chip (A) above is exactly the same colour as the centre square shown at (B) (shown below) using the tramlines (which are exactly the same colour) you can see how the squares [a] appear to change colour [b] even though they are all the same colour!

Thanks to Edward H Adelson who is Professor of Vision Science at MIT (USA) for discovering this very odd optical effect!

http://web.mit.edu/persci/people/adelson/checkershadow_proof.html

Dudeney's Cut-out 'SQUARE' 2

Another tribute to Henry Ernest Dudeney, an English mathematician who devised many challenging Maths shapes-to-fit puzzles like this one from 1917.

As before - this puzzle is based on rectangles split in half equally. (Width is half the height), you can Photocopy and carefully cut-out each shape and rearrange these shapes into a square box.

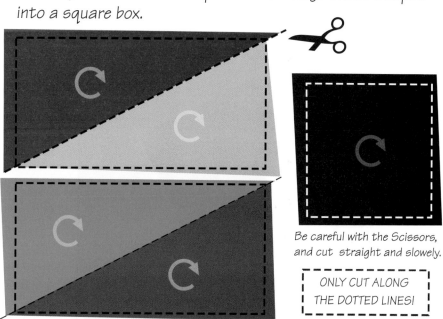

Be careful with the Scissors, and cut straight and slowely.

ONLY CUT ALONG THE DOTTED LINES!

It may be helpful to know that this is fairly easy to solve in that you only need to rotate the four Triangles and square to make the correct box shape. Do not 'flip them' i.e. you do not need to turn them over.

We shall be featuring more of Dudeney's geometric puzzles - but not all are as easy to solve as this one! The answer to this puzzle is overleaf but give yourself 5 minutes or more...

Dudeney's Cut-out 'SQUARE' 2

Did you solve this puzzle easily?
All you needed to do was rearrange the Triangles and square into a box shape, like this!

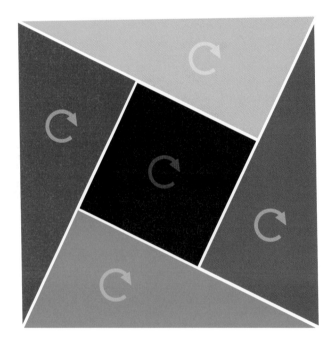

Credit: This puzzle is attributed to the work of Henry Ernest Dudeney (Amusements in Mathematics - 1917)

It's a neat way to solve the problem. **And a fun way to explore Maths - using just geometric shapes.**

Both Lewis Carroll and Henry Dudeney had fun playing with odd shapes and creating puzzles and solutions from the world around them. We shall be looking at some more Dudeney and Carroll puzzles in our other pages.

Match Puzzle: **From these six matches make four triangles**

To solve this you need to explore shapes that you can make into Triangles...

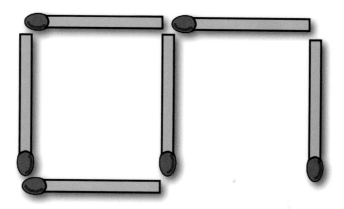

Not impossible! It should take no more than 5 minutes playing with sticks, pens or pieces of card. It may help to 'draw' a shape to represent Triangles before you start.

Sometimes you can see the shape immediately (imagine it) and other times it's not obvious until someone sees it before you!

Don't turnover page unless you are 100% sure you are correct, (it's too easy).

Match Puzzle: From these six matches make four triangles

It's not a trick question, you just have to think different as in 3 dimensions!

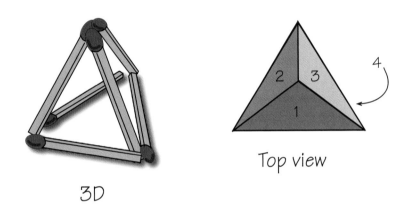

3D

Top view

Well, we did say it was easy when you think 3D.

Your next challenge is to make 9 triangles with just twelve matches, (think 3D as we did).

Create your own puzzles (based on this 3D model) and see how you can make triangles and shapes easily.

Try another shape with a six sided Hexagon shape. All are possible when you-know-how with matches and a bit of plasticene, but 'Geomag' magnets are our favourites for trying out different shapes like this.*

* https://www.geomagworld.com/en/

What are the 'LIGHTEST' and 'HEAVIEST' materials in the world?

What actually makes things 'heavy' has been puzzling Scientists for some time... You may be surprised to learn that some metals are lighter than water and some metals are heavier then Lead. Gold is surprisingly 'heavy'. There could be 'other' materials which are be both lighter or heavier - as yet to be discovered.

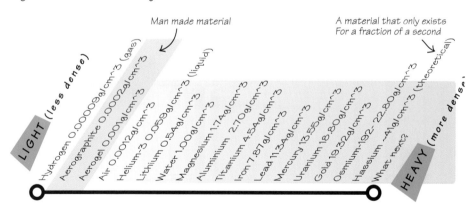

Man made material

A material that only exists For a fraction of a second

LIGHT (less dense)

Hydrogen 0.00009g/cm^3 (gas)
Aerographite 0.0002g/cm^3
Aerogel 0.001g/cm^3
Air 0.0012g/cm^3
Helium-3 0.059g/cm^3
Lithium 0.534g/cm^3 (liquid)
Water 1.00g/cm^3
Magnesium 1.74g/cm^3
Aluminium 2.70g/cm^3
Titanium 4.54g/cm^3
Iron 7.87g/cm^3
Lead 11.34g/cm^3
Mercury 13.55g/cm^3
Uranium 18.80g/cm^3
Gold 19.32g/cm^3
Osmium 192.- 22.80g/cm^3
Hassium ~41g/cm^3 (theoretical)
What next?

HEAVY (more dense)

As you can see Hydrogen gas is the overall lightest with Gold and Osmium as being some of the heaviest metals you can find. Note: that being simply 'dense' does not make for a 'hard' metal i.e. Gold and Lead are both 'heavy' yet also relatively soft metals. (Whilst Mercury is a liquid).

See over and discover if these particular properties as 'weight' compare.

Are these the 'LIGHTEST' and 'DENSEST' materials in the world?

Although you cannot get anything lighter than Hydrogen (other than being in an absolute vacuum such as Space). There is a theoretical limit to density or weight. We know (thanks to Isaac Newton's second law of motion) that 'Weight' (as a force) is the 'Mass' (size (volume) and material density of the object) times the 'Gravity' (w=mg). Gravity is almost constant on the Earth's surface. So the weight comparison would be the same (even in outer space)*.

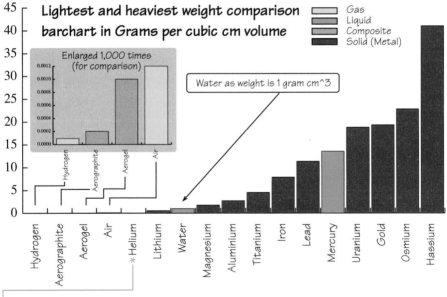

Lightest and heaviest weight comparison barchart in Grams per cubic cm volume

Gas
Liquid
Composite
Solid (Metal)

Enlarged 1,000 times (for comparison)

Water as weight is 1 gram cm^3

* Liquid Helium can only be liquid at -270 degrees C (only 3.19 degrees above absolute zero). So this is NOT like the rest of the materials.)

*If it has weight on Earth that would remain the same for comparison, even in outer space. Newton's second law would still apply to describe weight although Weight and Mass are different. And yes weight would vary on another planet such as Neptune and Mass would increase with speed according to Einstein.

Find out more about Issac Newton's three laws of Physics:
http://simple.wikipedia.org/wiki/Newton%27s_laws_of_motion

Euler (trail) circuits

Firstly the rules are very simple. Just 'join-the-dots' in a continuous line once without overlap!

For this puzzle we are going to use Pencil and paper and draw this basic circuit on a scrap of paper. It does not have to be accurate, just close enough will do.

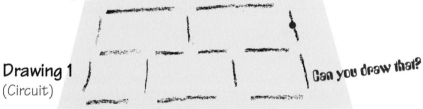

Drawing 1
(Circuit)

Can you draw that?

Using a pencil (or pen) join each line circuit section (above) with a continuous line and join them all up without crossing any lines. To get you started we have drawn a RED DOT to connect up with all the other sections.

Drawing 2
(Envelope)

Is it easy or hard?

This is an envelope shape and you have to try to connect each dot (called a NODE) joining all the dots without the pen leaving the paper.

Drawing 3
(Mess)

Have a go!

This is a mess. It is a horrible mess but if you can count how many RED DOTS there are then you can predict if they can all be connected without getting your wires crossed!

The answers to all this simple deception is overleaf.

Euler (trail) ⌃ circuits

CIRCUIT RULES

1. A Euler (or Trail) circuit joins every edge (or line or dot) only once.
2. An intersection (of lines) is called a NODE
3. An intersection can have any number of ODD or EVEN lines.
4. A NODE is either ODD or EVEN depending on intersections.
5. You can ignore particular NODES that have an equal number of lines at intersections (as shown in diagram 2).

You can quickly work out what can be done and what cannot be done by counting the NODES. This is the tricky part but if you can get that right you can make a quick calculation on IF the NODES are even or not. So yes it can be done IF there are an ODD number. If the NODES are ODD numbers then it can be done, but you still have to find out how to do it. **And that can be a puzzle!**

Two examples:

Drawing 1 cannot be done (even number of NODES)

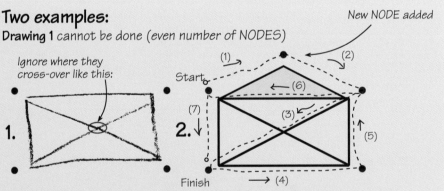

Drawing 2 The Envelope (above) cannot be done unless you add another NODE to make it an odd number (as shown on the Envelope above).
Drawing 3 can be done (it has 27 as an 'odd' number of NODES). That means it is possible to do. (But it does not show you how to do it!)

Eular Trail Puzzles can be great fun and there are many types of INTERACTIVE circuits to try out. Better to KNOW YOUR NODES > You might even become an expert one day. See this weblink below:

http://www.flashandmath.com/mathlets/discrete/graphtheory/euler.html

IF you could fold a flat piece of paper in half 50 times... and placed it on the ground how high would it be?

Try to imagine that our paper is a thickish 1mm, (normal paper might be 0.001mm) but even so if you could fold any wafer thin paper thickness **50 times** the result would be very surprsing!

Folding paper by one fold (by one each time)...			
0	**1mm**	28	13.4217728
2	2	29	26.8435456
3	4	30	53.6870912
4	8	31	107.374
5	16	32	214.748
6	32	33	429.496
7	64	34	858.992
8	**1.28cm**	35	1717.968
9	2.56	36	3,435.968
10	5.12	37	6,871.936
11	10.24	38	13,743.872
12	20.48	39	27,487.744
13	40.96	40	54,975.488
14	81.92	41	109,990.976
15	**1.638m**	42	219,901.952
16	3.2768	--	439,803.904
17	6.5536		(Distance in km to the Moon)
18	13.1072	**43**	**1.144 LD**
19	26.2144		(LD = Lunar distance)
20	52.4288	44	2.288
21	104.8576	45	4.576
22	209.7152	46	9.152
23	419.4304	47	18.304
24	838.8608	48	36.608
25	**1.6777216km**	49	73.216
26	3.3554432	50	One more step!
27	6.7108864		

To put these distances into context of space travel...
384,400 is the average distance to the Moon!
If you look at the last but one (no.49) figure of
73,216 LD - this is 28,144,230.4km (kilometres).
i.e. which is ONE FIFTH of the way to the Sun!

Cryptic Mini Puzzle

Can you read this?

I cnduo't bvleiee taht I culod aulaclty uesdtannrd waht I was rdnaieg. Unisg the icnndeblire pweor of the hmuan mnid, aocdcrnig to rseescrah at Cmabrigde Uinervtisy, it dseno't mttaer in waht oderr the lterets in a wrod are, the olny irpoamtnt tihng is taht the frsit and lsat ltteer be in the rhgit pclae. The rset can be a taotl mses and you can sitll raed it whoutit a pboerlm. Tihs is bucseae the huamn mnid deos not raed ervey ltteer by istlef, but the word as a wlohe. Aaznmig, huh? Yaeh and I awlyas tghhuot slelinpg was ipmorantt! *See if your fdreins can raed tihs too!*

Credit: Cambridge University Research

Note: all spelling deliberate. The way to read this is a demonstration of how our word recognition works when we speed read, if we already know the actual English spelling then the meaning is resolved automatically as we read. It's important that the 'first' and 'last' letters are in the right place...

Logic Button problem

Only one button stops the launch of a deadly virus, but each button has a false (or truthful) message attached. *Which one is correct?*

says GREEN is NOT the real button

says BLUE is NOT the real button

says both RED and GREEN are NOT either!

A puzzling problem. Which button to press to stop the virus attack?

Notes:

This puzzle presumes two buttons are hiding something then we need to work out which is the 'true' button to press to save us from the virus.

We do not need any more information. This is all we have to work on (and it is possible). **Think positive!**

Logic Button solution

This puzzle is based on the prior logic maths of Lewis Carroll who invented a lot of similar problems which can be solved in the same way... *using logic.*

So before I show you how to do it you might like to read 'Alice in Wonderland' and wonder how a Mathematician could write such an imaginative book. Or make such impossible puzzles that seem impossible to solve...

If you have already read Alice in Wonderland perhaps you can solve this tricky problem of how Alice make make sense of what she is told?

RED button - says Green is WRONG (but could be lying about that).

GREEN button - says Blue is WRONG (but could also be lying)

BLUE button - says Red and Green are both lying!

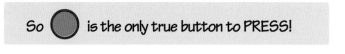

So   is the only true button to PRESS!

But how did I know that? Is it logical?

Only one button can be the true answer. As Blue cannot state that both the 'other' colours are false without lying, we can exclude BLUE. That leaves RED or GREEN buttons. If we consider that all three buttons are actually telling the truth then two are telling 'lies'. This is our clue. As RED states that Green is wrong. BLUE does not state Green is wrong. BLUE cannot be right about both buttons - therefore that just leaves the GREEN button... we are saved!

But of course you knew that all along! Try out some more word logical puzzles on this web link below and see how you can become your own 'Alice in Wonderland' logic sleuth:

What makes this *Lemonade* so tasty?

Most lemonade bottles contain virtually no Lemons at all (at best 2% or four tiny teaspoons of Lemon juice). So what is the supermarket Lemonade actually made of?

SUGAR is not considered 'sweet-enough' for your 'fizzy' Lemonade so they add lots of artificial 'sweeteners', replace sugar, i.e. Aspartame is x180 times stronger than sugar). Next it's 'fizzy water' and that is the real key ingredient: Carbon Dioxide (CO_2) 'pressurised in the plastic bottle (at about 60 psi*), it tastes of bitter Lemons (as it fizzes on your tongue). **It's a *Carbon Dioxide gas!***

2%
Lemon juice!

One Litre = 1000 ml

You only get this much (2%) Lemon juice AND ONLY if it that is on the label...

*The plastic bottle is gas pressurised at about 60 pounds-per-square-inch, the plastic bottle is safe for 100 psi (or more) but never used in case of explosion!

Making a 'dried Lemonade' recipe:

It's really a 'Sherbert' recipe and will taste exactly the same as Lemonade does. (It will still rot your teeth, if you don't clean your teeth afterwards) - just like the shop Lemonade does!

Mix the following in a cereal bowl:

a) 4 egg cups of ICING SUGAR (used in cake making)
b) 1 egg cup of BAKING SODA (used in cake making)
c) 1 egg cup of CITRIC ACID (used in bread making)
d) 1 egg cup of TARTARIC ACID (used in cake making)

Mix and TASTE until you think it is right. The fizzy lemonade 'taste' comes from the CARBON DIOXIDE bubbles that 'explode' when the Citric Acid (acid: ph 0-6) and the Baking Soda (alkali: pH 8-14) are mixed on your tongue (creating sodium citrate) . The Tartaric Acid just adds a smoother taste, whilst the Sugar makes it a lot sweeter.

Carbon bubbles are totally harmless. Artificial Sugar often has no 'food value' at all hence 'zero-calorie' health drinks and foodstuffs are sold as 'healthy' alternatives... (see overleaf)

Discover more about *Carbon Dioxide* in fizzy drinks:

http://www.buzzle.com/articles/carbon-dioxide-in-soda.html

More puzzling NO SUGAR! *Lemonade!*

Lemonade is mostly Carbon Dioxide bubbles. *It's the 'taste' of these bubbles which appeals to our taste buds. Carbonated Mineral water was (and still is) sold by J. J. Schweppes who developed further the process of 'artificially carbonated water' discovered by the English Chemist Joseph Priestley*. He studied gaseous liquids as a by-product of the fermentation process such as in 'beer brewing' which uses 'natural sugars'.*

What are 'natural sugars' made of?

Breathe in and out. It's $C_{12}H_{22}O_{11}$ = which means 12 Carbon atoms, 22 Hydrogen atoms and 11 Oxygen atoms all mixed together to form sugar - also called sucrose - (table sugar).

What is **replacing 'Sugar'** in your Lemonade?

There are lots of things sweeter than Sugar, **(SUGAR = 1)** - so here is our **top 10...**

Rank	Name	How much sweeter?	Natural
10	Pentadin	x500	Yes
09	Curculin	x550	Yes
08	Sucralose	x600	No
07	Brazzein	x800	Yes
06	Neohesperidin dihydrochalcone	x1500*	No
05	Thaumatin	x2000	Yes
04	Alitame	x2000	No
03	Monellin	x3000	Yes
02	P-4000	x4000	No
01	Neotame	x8000ß	No

DANGER Beyond x1000 'sweeter' you have to wear a face mask when handling this substance as it will result in a severe headache or you'll feel very ill afterwards.

Note: Saccharin is x300 and Aspartame is x180, (both common sweeteners in 'soft' drinks). For this reason DIET cans 'float' on water (less dense) and sugar based cans should sink!

'E'-numbers are usually 'preservatives' or 'sweeteners.'

If something contains 'NO SUGAR', it usually contains artificial sweeteners. If it has E numbers e.g. E458 type the 'e' number into a search engine and it will (usually) be either a preservative or a sweetener. Not all 'E' numbers are harmful. (as even Oxygen has an E number), but you'll find more 'sweeteners' there. http://en.wikipedia.org/wiki/Sugar_substitute

FLOAT

SINK

* http://www.rsc.org/diversity/175-faces/all-faces/joseph-priestley

Find my missing Yogurt...

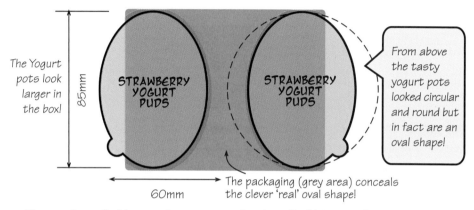

The Yogurt pots look larger in the box!

85mm

STRAWBERRY YOGURT PUDS

STRAWBERRY YOGURT PUDS

From above the tasty yogurt pots looked circular and round but in fact are an oval shape!

60mm

The packaging (grey area) conceals the clever 'real' oval shape!

The author of this puzzle was surprised to find his new favourite yogurt pot was not 'ROUND' but in fact an 'OVAL' shape concealed inside the wrapper. So presuming the pots were originally round to begin with (like most yogurt pots), you will (hopefully) find my missing yogurt!

I shall have to presume that round Pots and the oval shape are of the same depth, otherwise any comparison is pointless!

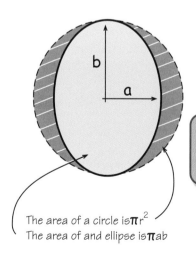

b

a

We know the pot width is 60mm and height 85mm as an oval and we know that the round pot size would be 85mm diameter (we only need the radius to work out the area as shown a= 30, b=42.5).

So the question is; what percentage of the Yogurt pot area has changed if it had been round instead of an oval shape?

The area of a circle is πr^2
The area of and ellipse is πab

You can use the formula on the left shown. Subtract one from the other and that will show how much yogurt is missing.

Which yogurt pot would you choose? Answers to the puzzle overleaf.

Missing Yogurt -puzzle...

I think I solved it... This may be the solution...

The area of a circle is πr^2
The area of an ellipse is πab

The area of a circle is therefore:
3.14159 x radius 42.5^2 (half of 85mm) which is 5674
(rounded up to nearest whole number)

Subtract:
The area of the ellipse is πab is:
3.14159 x 42.5 x 30 which is 4005 (rounded up to
nearest whole number)

So a percentage difference would be:
Circle area divide Oval area as a percentage:
4005 ÷ 5674 = 0.705 x 100% which as a percentage
(70.585125% approx. **say 70%**)

Now take 70% away from 100% = **30%**

**No wonder I am still hungry, almost a third of my yogurt pot
has completely disappeared even before I have tasted it!**

*The missing yogurt puzzle has been solved. I still miss my
30% but then perhaps I eat too much yogurt!*

Make a **5** piece SQUARE!

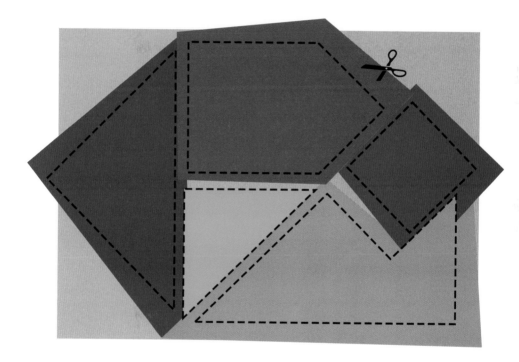

Your task is to construct a perfect SQUARE from these 5 pieces of paper. Of course you can 'reverse' (use EITHER side) of the pieces should you need inspiration. There also may be more than one way to do it.

Ideas: You can trace this page and colour it in (both sides) or photocopy and stick the two sides together or (easier still) ; use any coloured card you have and then cut out the (dotted line) shapes. **Make a square if you can!**

ANSWERS FOR a 5 piece SQUARE!

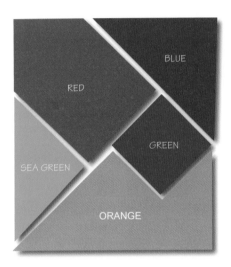

OR:

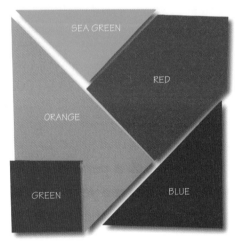

If you managed to get either of these answers you
have done well to solve this puzzle. As far as we know there
are no others that can be made.

What happens to Water and Ice in a **Microwave?**

Here is an experiment that will confound your parents (and possibly) your Teacher!

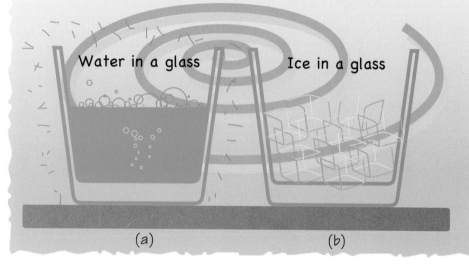

Water in a glass   Ice in a glass

(a)   (b)

1. TAKE TWO GLASSES and fill one (a) with normal tap water and the other (b) with some VERY COLD FROZEN ICE CUBES.

2. Place both on the rotating Microwave dish (without glasses touching each other) and then put them on full power 'cook' for about 2 mins.

SO - what do you think happens next ?

Many would say it's obvious! But this simple experiment - is not quite as clear - or as easy to understand as you think!

Note: Danger - do not ever PLAY with Microwave ovens at home.

> This is an experiment to do with your parents or Teacher only OK!

Discover what happens overleaf, you may be surprised!

Answer: they don't melt (at first)!

The scientific principle is that Microwaves are directly related to RADAR technology (as in radio frequency). It was discovered in the Second World war that unfortunate 'accidents' could happen if you stood in the path of a Radar beam. You could get badly cooked and it was (then) a mystery.

How it worked came much later but Scientists did discover that Microwaves do excite the molecules in water - that heats food. It is the water molecules (molecules rotating at microwave frequency) in our food (or drink) that actually heat up (not the food itself (as it needs water in the food). Don't stand too close observing this, the wire mesh door is your ONLY protection from a serious burn!

The Puzzling part is that they don't melt because...

As ICE is solid - the radio waves go straight through the ice as it does through the Glass beaker. IF however the ice starts to melt in warm air temperature - even a little liquid will quickly heat up quite fast just like the water in the glass does... It can get HOT, very quickly so be careful.

If you would like to find out who discovered Microwaves (or even RADAR) then take a look at the Wikipedia entry on Microwaves here:

http://en.wikipedia.org/wiki/Microwave_oven

The outer square 'size'

The aim of this puzzle is to find the perimeter of the larger square (without measuring it)

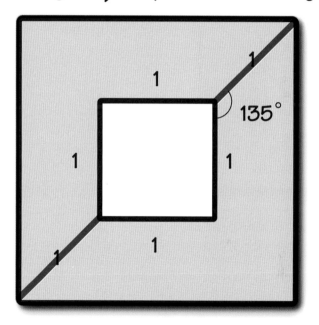

Note: this is NOT a scale drawing

Clues:

(a) All the measurements are one unit. **You can use a Calculator but <u>NOT</u> a Ruler, a Protractor or a Compass!**

(b) All the corners are square. **i.e. at right angles to each other.** The thickness of line (or colour) is not important.

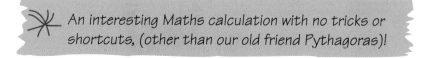

An interesting Maths calculation with no tricks or shortcuts, (other than our old friend Pythagoras)!

Can this be solved with geometry or basic Maths? *Answers overleaf.*

The outer square 'size'

The answer is revealed - using only Pythagoras

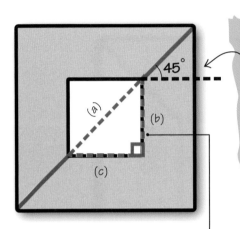

45°

(a)

(b)

(c)

Any square diagonal must be 45 degrees (although the original measurement shows 135 degrees - that is because 180 degrees minus 45 degrees = 135 degrees). We can also see that this is a right angled triangle (in the smaller square left) and we can use Pythagoras Theorem ($a^2+b^2=c^2$) to work out the longest side length 1 (the hypotenuse).

1 SMALL SQUARE | smaller triangle |
Using Pythagoras theorem: ($a^2+b^2=c^2$) where (a) is simply (square root of two) as both (b) and (c) are of one unit each they can be eliminated here. **Therefore: the diagonal of the small square is simply = sqrt2**

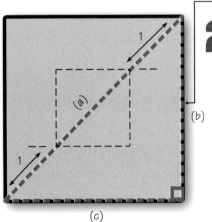

1

(a)

(b)

1

(c)

2 BIG SQUARE | larger triangle |
We now add the other TWO hypotenuse - (of 1 unit each) which we include below:

$$2a^2 = (2+\textbf{sqrt2})^2$$
$$a^2 = 2+(sqrt2)^2$$
$$a = 2 +(sqrt2)/(sqrt2)$$

transposing to:

$$(sqrt2)+2/2 = 1+(sqrt2)$$

times four (the sides of the square):

$$4(1+sqrt2)= 9.66$$

The overall large square perimeter is therefore 9.66*

*(to 3 decimal places)

How many Shreddies are there in THIS box?

> **This is a really BIG box (1Kg) - but it made me think - just how many Shreddies is that?**

It's a bit of a puzzle but we do have TWO good clues on the front of the box...

 + **Per 40g serving with 125ml semi-skimmed milk**

This does not indicate how many SHREDDIES there are actually in the cereal box - but we do know the 'weight' of the 'serving' in the bowl (and it includes semi-skimmed milk)

This makes a rather good puzzle to try and work out!

There may be more than one way to work it out. I decided to do some research online to see how 'skimmed-milk' is measured by weight (in grams). But first a quick calculation:

Per 40g serving with 125ml semi-skimmed milk **=** *Per 100ml skimmed-milk serving. I divided 125 by 100 for a factor of 1.25. Next I divided the 40g by 1.25 to give me the 100ml serving.* **=** **Per 32g serving with 100ml semi-skimmed milk**

➤ *Do NOT try to weigh an individual Shreddies on your Mum's kitchen scales! That's just cheating. There is another interesting way of finding how many Shreddie there are in a 1,000 gram box by learning a little about how to weigh milk . See next page to find out!*

You **don't** have to count them all!

If there are 25 servings per box (it states) we can also see that if we divide 1,000g by 25 = 40 grams (with semi-skimmed milk).

SKIMMED MILK

We need to know exactly how heavy the skimmed milk is...

Milk is measured according to the UK Ministry of Agriculture, Fisheries and Food; Food Portion Sizes, HMSO (1988)]. The average portions of milk are based on nutrient values given per 100mls, as per 100g serving The specific gravities (densities) used to calculate the volume are: Whole milk 1.04, Semi-skimmed milk 1.03. (**Note:** 1.03 is the same as 1.03%)

▶ Note: Milk density is compared to 100ml of Water. If you divide 1,000 grams by 100 millilitres you will get 10 grams. As almost all milk is 'standardised' We find 100ml skimmed milk is 1.03 which makes 100ml = 10.3 grams weight.

http://www.milk.co.uk/page.aspx?intPageID=43

The weight of the Shreddies box is 1Kg (1,000 grams) as shown on the box. The weight of a bowl of Shreddies is therefore 29.7 grams, because 40g of Shreddies minus 10.3g of semi-skimmed milk is 29.7g.

We still don't know how many **Shreddies** are in a box!

So we are going to guess! You're not supposed to guess in Maths but I can see no other way!

I have to guess the weight of ONE Shreddie (with no milk). **I think it is 0.5 grams.**

By guessing the weight of one Shreddie in a spoon we can use that to divide into the 'standard' weight of a bowl of Shreddies. (That is what we did above - subtracted the 'standard' weight of the semi-skimmed milk).

This what we do know. *Shreddies in a bowl: 29.7g (without semi-skimmed milk), Shreddies x25 servings in a box. So 29.7 x 25 = 742.50 grams per box is NOT the right answer - as we know it should be 1,000 grams (1Kg). So my 'guess' at the single Shreddie weight is far too small. But from that we can now work out what it should be. It is 34% larger than 0.5g. (Divide 742.50 into 1,000 and multiply by 100 = 34%). So 0.67g is a more accurate weight for one Shreddie. Simply dividing that into the weight of the box will give us the answer.*

So the number of Shreddies in a BIG box of Shreddies is 1,492 approximately!

For more interesting observations on Mass and Weight, see **'A kilogram is about'** and find that its , not just about cereals. **https://www.mathsisfun.com/measure/metric-mass.html**

What if we were to 'CUT-OFF' the top half of an isosceles Triangle?

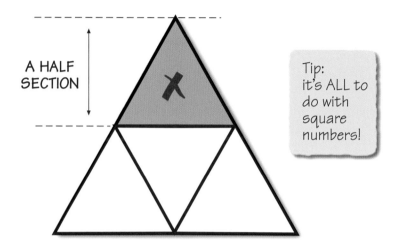

A HALF
SECTION

Tip:
it's ALL to
do with
square
numbers!

FOR A HALF - we could end up like this and the answer here is simply a quarter as you can see, there are four triangles above.

NOW try this slightly harder version. If we wanted just a third of a triangle. How would we do that?

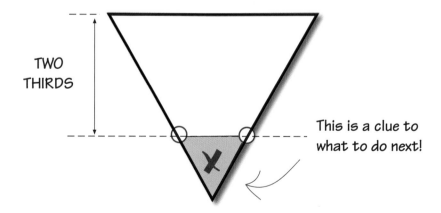

TWO
THIRDS

This is a clue to
what to do next!

Although the answer is overleaf you might like to try to solve it first.

You can see that **THIS IS EASY!**, just make even <u>more</u> triangles!

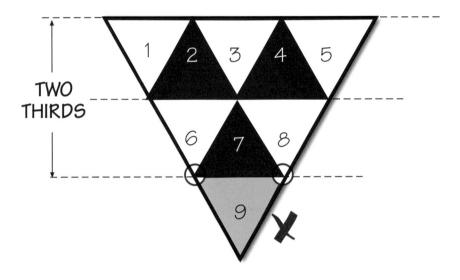

(and the answer above is a ninth!)

Which is a square number (3 squared). You may remember square numbers (see page 69) are in PASCAL's triangle as well. We can work out that we can create any number of smaller or larger Triangles simply by using square numbers...

Plus if we knew of the measurements we could also work out the area of this Isosceles Triangle (or any part) using Heron of Alexandria's* rather clever method of calculation).

*That is hidden away on page 139 of this book!

HUMPHRY DAVY AND HIS MAGIC LAMP!

This may may not look much of a 'magic lamp' but it has saved thousands of coal-miners' lives all over the world since its invention in 1815.

The 'Safety lamp' was invented by Sir Humphry Davy in 1815. It consisted of a glass cylinder within which the flame was further encased in wire gauze so as to permit air to enter but prevent the flame escaping to ignite any inflammable gases which might be present in the invisible mine air.

Puzzle solved! Deep Coal mining released explosive gases (often Methane) that naturally occur below ground. Often these are natural gases (like your gas cooker which has a strong smell 'added' because it has 'no-smell' at all). Such gases are highly explosive in confined spaces and set alight by any nearby flame (or candle) that the miners used. Many died. This was a huge problem for the new industrial age which relied on coal completely...

Humphrey Davy was able to demonstrate his safety device at the Royal Institute in London in 1815. He was already a famous Chemist and went on to discover several important elements in the periodic table.

Davey could show with a simple experiment how effective his Magic 'Safety Lamp' was to save lives...

It became the 'Davy Safety Lamp' and was quickly copied all over the world and became a true life saver. Miners also reported an additional feature: it was able to tell what kind of gas was present in the mine by the colour of the flame in the lamp. So sensitive is the flame to a gas that it is still a safety requirement to have one today - even with modern electric lights and torches.

http://www.nahste.ac.uk/isaar/GB_0237_NAHSTE_P0175.html

IT'S NOT MAGIC IT'S SCIENCE...

Two simple experiments to test if Humphry Davy was right... <u>WARNING do not do this without supervision</u> - a bit pointless demonstrating a safety lamp if you catch your mum's curtains on fire!

> You will need a TEA-CANDLE, SAUCER, GAUZE sieve (or TEA-STRAINER) and a box of safety matches.

1.   **2.**

1. First experiment is to light your candle and carefully place your gauze over the top.

Humphry Davy could show that the flame refuses to pass the wire mesh.

The wire mesh allows the light to escape but not the heat, which could cause a 'gas explosion' in a mine shaft or pit.

2. This second experiment was NOT part of the Humphry Davy demonstration.

First light the smoking candle again and then immediately extinguish it. Wait a second or two then with your lit match catch the smoke 3 - 4 inches above the candle and you will see the flame 'jump' down to the candle. This shows how dangerous forest fires can be. The flame can 'jump' from tree to tree, even if the fire appears to go out!

The Magic Lamp cannot prevent 'fire jumps' from occurring in Forest fires.
http://www.nhm.ac.uk/nature-online/earth/volcanoes-earthquakes/forest-fire/

Q. All these **Roman Clocks** have something WRONG in common - and it's <u>not</u> the time!

So what could that 'something' be?

You will probably have a Roman clockface at home or at School - even a Wristwatch will have the same clockface problem. It is very hard to spot because you see it every day and have never noticed it!

And you will probably NEVER guess what it is....

Discover what makes these Clocks 'tick' overleaf:

A. Those **Roman Clocks** do have something **WRONG** in common...

It is in the Roman numerals:

I, II, III, IV, V, VI, VII, VIII, IX, X, XII and XII

You may have noticed that ONE of the numerals (on the clockfaces overleaf) is wrong. **There is NO Roman numeral that uses 'IIII' (and that's the puzzle).** This was a deliberate 'falsie' by the French Carriage Clockmakers to **'balance'** the other side of the clockface **visual 'weight' of the number VII.**

There is no logic to it (other than that it is widely accepted) and nobody wants to change it. There is however some dispute about how Roman numerals are used as there seems to be some wide variation, even by the Romans themselves.

GMT are undecided why Roman clockfaces do not use IV
http://wwp.greenwichmeantime.com/info/roman.htm

The 'Mpemba' effect

A schoolboy noticed that WARM WATER freezes faster than cold water, and nobody really knows why!

(1) Get two jugs and Thermometer(s).

(2) Make one about 40 degree C and the other 10 degree C

(3) Put them in the fridge freezer cabinet

(4) Every 5 mins measure the temperature (and write it down)

(5) Gasp as the warm/hot water freezes faster into ICE!

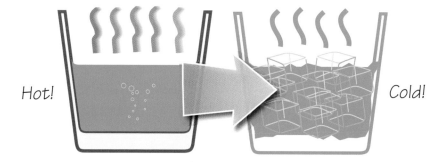

Hot!   Cold!

How does this happen?

Nobody is sure why, it's all very puzzling, many have tried to explain this effect - but all their 'findings' are inconclusive.

Why is it called the Mpemba effect?

The effect was discovered in Africa by Erasto B. Mpemba. In 1963. He was ridiculed by his classmates and even his Science teacher! Eventually he told this to a professor who then carried out the same experiment to find that he (Erasto) was right. The professor, with Mr Mpemba, jointly published their findings. Probably with much embarrassment to his Science teacher who thought it impossible!

http://www.lancs.ac.uk/ug/thompsom/

More strange properties of water!

Water has other unique properties that are often just as baffling. And just as Mpemba found, there are more surpises to be found. Here are our top 10 water puzzles.

True or False?

[1]   Water does not (always) boils at 100 degrees C

[2]   Water does not (always) freeze at 0 degrees C

[3]   Water (uniquely) expands on freezing (so it floats on water)

[4]   Water can melt under pressure (so you can ice-skate)

[5]   Water can 'shrink' by a (small) volume when it is mixed with spirit

[6]   Water is the world's best universal 'solvent' able to dissolve most natural substances i.e. Salt

[7]   Water is inert (not flammable) but Oxygen and Hydrogen are highly explosive mixtures (as used in Rocket fuel).

[8]   Water is 'by product' when rockets burn Hydrogen and Oxygen

[9]   Water is the basis of all life (and weather) on Earth

[10] Water temperature and salinity in the sea varies (and so limits the weight of any cargo ship travelling across the oceans).

Perhaps you too will find 'other' unique properties of water: some of which have been found only by accident (as Erasto Mpemba did).

Look at a glass of water and wonder what it is, how does ICE freeze faster in warm water - rather than cold? Your mission is to find if all the above is true or false. Some answers may be 'easy' others are (as Mpembo found) not as you might expect!

http://www.lsbu.ac.uk/water/anmlies.html

The Triangular puzzle

What is the perimeter of the triangle?

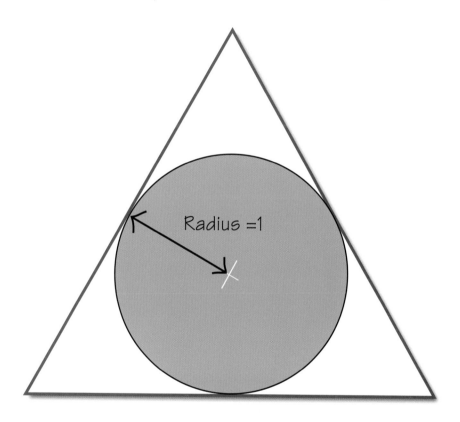

Radius =1

The only measurement you have is the radius of the circle. *Rulers are forbidden, calculator with sine, cosine, or tangents. It can be done using simple geometry and pure logic solving.*

See if you can find a better way than ours before you look at the NEXT page!

The puzzle solved...

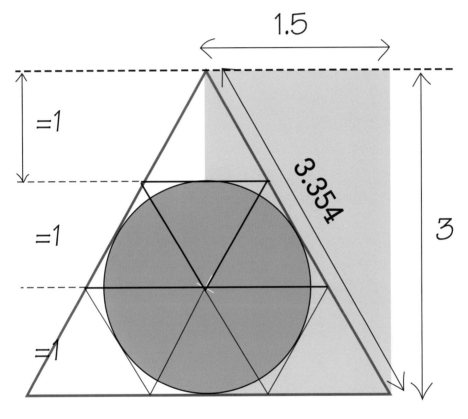

circle radius =1

Therefore perimeter = 10.062

You'll note that you can split the triangle into 9 equal smaller Triangles. Simple geometry and Maths solves the rest.

Pendulum Swings...
How long does a pendulum swing for?

You can observe this with a piece of string and a washer tied to the end of a weight.

Notice: that the longer string takes more time than the shorter string to swing to and fro. A bit like a clock mechanism, so we can calculate the time it takes by using this formula...

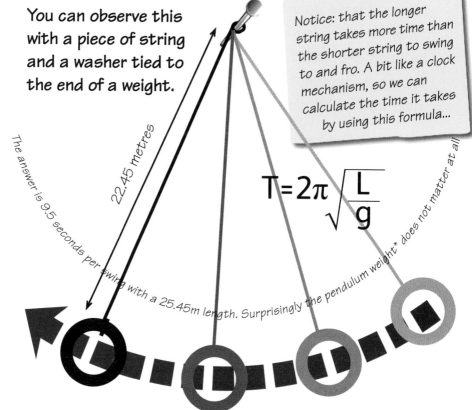

22.45 metres

The answer is 9.5 seconds per swing with a 25.45m length. Surprisingly the pendulum weight* does not matter at all

$$T = 2\pi \sqrt{\frac{L}{g}}$$

As you would expect someone has worked it out for us already!
The formula above (or equation) is all that we need know to find out what the T, L and G stand for below - and it's not that difficult! (See below) :

T is the TIME in seconds
pi = 3.14 as in the Greek letter π
SQRT as in the SQUARE ROOT symbol $\sqrt{}$
L is the actual LENGTH of the string in metres (or feet)
g is acceleration due to GRAVITY (9.8 m/s - or 32 ft/s)*

---

*Pendulum swings are not affected by weight but by length of the pendulum swing and the gravity (which is 9.8m/sec² - the acceleration due to Earth's gravity). A one second time requires a length of 1m swinging to and fro. The Science Museum Focault pendulum is 22.45m high. Therefore the time period is 9.5 seconds precisely.

Is that the 'puzzle', a clock?

This clock also measures other things found 'puzzling'...

You may have seen the Foucaults pendulum hanging in London's Science Museum. Although it was not 'invented' by Jean Bernard Leon Foucault - he is credited with the first demonstration of a large swinging pendulum, held at the 'Worlds Fair' at London in 1851....

What Foucault had shown in this demonstration was that the Earth's rotation could be seen as the pendulum swung - it moved by 0.2 degrees every minute. (x60 x 0.2 = 12 hours)

The Earth's rotation would alter the pendulum swing - as it moved by 0.2 degrees every minute. Earths GRAVITY does not change that much but it would be a different value on another planet (like the Moon or Jupiter)...

As you now know from the previous page, the length of string, or the weight, does not make any difference. It is affected by gravity (which does not change much around the world).

Foucaults pendulum was not really about measuring time. It was made to prove that the world was spinning and it could be measured accurately. The weight at the end of the pendulum moved slowly in a big circle as it moved back and forth and that is why it had to be hung at the highest point in the building. The one hanging in the **Science Museum** is one of several across the world. You can still see it working although some people still puzzle over how it moves.

http://www.animations.physics.unsw.edu.au/jw/foucault_pendulum.html#1

Polaroid puzzling vision

You'll need TWO pairs of Polaroid sunglasses and a school protractor.
You can also change the Protractor for anything
clear and flexible - such as a clear plastic ruler.

Stage one. The first puzzle is to find some Polaroid specs. Many sunglasses look identical so this is your first challange.

Stage two. With the specs aligned (as above) at 90 degrees to each other) the top lens should go dark as you rotate it over the other pair.

Stage three. This is the interesting bit. Now the lenses are totally black you then sandwich a clear plastic protractor between the two lenses... and then you will see the distortions clearly in the protractor. You may have to TWIST it to see anything.

What's happening here?

Polaroid 'anti-glare' sunglasses work like minature 'sunblinds' - by filtering out unwanted light reflections. The Polaroid lens block the light that is not horizontally aligned (as light travels in particle waves normally being scattered in all directions). By rotating either pair of sun glasses you should see something.

Weblink: http://wonderworks.questacon.edu.au/polarised.html

Polarising Effects

We use two POLAROID sunglasses because of these unique 'anti-glare' properties but it is easier to use two small plastic polarised film sheets, (if available from school lab) one lens over the top. It works exactly the same way.

We can now do some interesting experiments...

[1] With one lens over another twisting one lens will change the strength of blocking the light. If you imagine each lens as having horizontal blinds (invisible to the naked eye) you can see how the Polaroid sunglasses work.

[2] Each lens has 'horizontal' blinds which block out the sun's glare but we can use them to see some interesting 'Polaroid effects' as the 'blinds are so small (micron level) they can break up the light and only allow light that is 'polarised' in one direction.

[3] In school by sandwiching the transparent plastic ruler (or school protractor) between the two polaroid sheets you can see a rainbow of colours appear. You can see 'stress' in a plastic material very easily (before it breaks). Use small sheets of Polaroid film (ask at School to experiment). And use your own clear plastic ruler in case it snaps!!

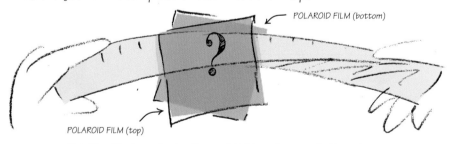

POLAROID FILM (bottom)

POLAROID FILM (top)

Twisting the top sheet will reveal hidden stresses in the ruler

Any old plastic ruler will do... and 'twist it,' (as if you're trying to break it)!
Just put one sheet above the ruler and the other below (the ruler), and twist!

Laser beams (by default) are already 'polarised' light beams. And by using a polarised (Polaroid) filter you can measure any 'rotation' of the beam with some accuracy in structures. Railways can be laid with far more accuracy using laser beams and polaroid filters. Other uses include your desktop or laptop LCD computer screen. Rotate some Polaroid specs (or polarised film) and see some puzzling results. **Interesting visual effects!**

The puzzling 65th square

How can it be that simply re-arranging these squares creates a totally new square?

[A] This is an $8 \times 8 = 64$ square (count them). Now re-arrange the shapes as shown in [B].

[A]

But if you should re-arrange the shapes:

[B] $5 \times 13 = \underline{65}$ square.

[B]

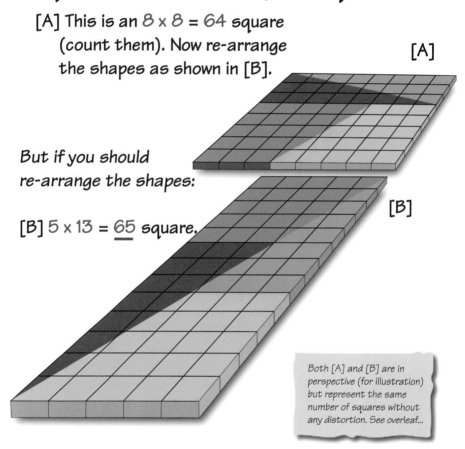

Both [A] and [B] are in perspective (for illustration) but represent the same number of squares without any distortion. See overleaf...

How do you explain that?

You may have an easy answer to this, it looks simple as it is based on classic Greek geometry puzzles but it never fails to amaze and inspire children. It is a challenge for teachers to explain how it is done!

squares?

How can you
be sure of this?

[A]

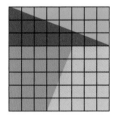

[B]

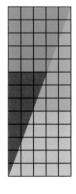

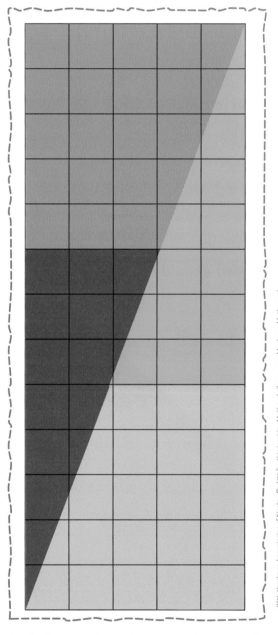

With thanks and permission of Kjartan Poskitt for this inspiring Maths challenge. www.MurderousMaths.co.uk

This is a 'hands-on' puzzle challenge.. First photocopy the page above and
then stick onto thin card. Then very carefully cut -out the FOUR COLOURED
TRIANGLES (shown inside the grey dotted lines). Follow and re-make the
diagrams to match diagram [A] and then [B] don't forget to count the
squares (on each) to see if you can solve the puzzle...

Take three **LEGO** bricks

All you have to do is take three LEGO bricks (A, B or C) and place them in either Box 1, 2 or 3 in sequence.

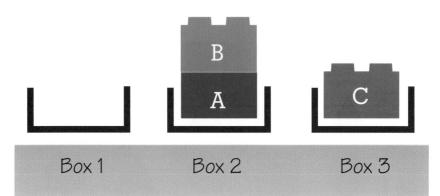

Box 1   Box 2   Box 3

The LEGO bricks can be placed in any position (in box 1, box 2 or box 3). But using the same bricks, same three trays, make another variation. You can have three bricks high in one tray, two bricks high, or just one brick per tray. You must use all three bricks (no more that three). How many variations do you think there are?

That is the Puzzle challenge!

Three boxes and three colours with three possible positions. Should be easy if you have a Lego kit!

Check your answers (with ours) on the next page...

That **LEGO** brick problem!

Well that should be easy, (not). So we had a few failed attempts. And then there is probably more than one way to do it. An easier way perhaps?

We decided to change the Lego shape (it did help our VARIATIONS sequence below). We didn't need Lego but it might have helped when we were stuck.

VARIATIONS: Starting with block a we changed each LEGO brick in turn until there were no other colour variations vertically.

POSITIONS: Starting with first block [a] we can ignore the VARIATIONS (above) and work out the positions for each Lego blocks in each box.

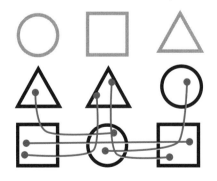

(1) The positions (marked in red) are limited to being two blocks high and using one other position. That limits the combination to 10 positions (we do not have to consider three blocks high as that would be included in our colour VARIATIONS above).

The Three Boxes can each (in turn) have three VARIATIONS. That makes 6. Then we add possible POSITIONS and that makes 5 and that doubles when you swop coloured brick positions making: 10x6=60. **We think we could make 60 in total.**

Join-the-dots...

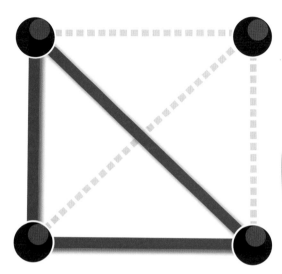

Create as many possible links as you can using just THREE strokes of the pen on paper...

Firstly the drawing on the LEFT is wrong...

Because: one last dot still stands (not connected).

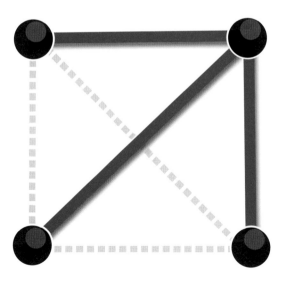

Correct!

This example (on the left) is just the start you need to join all the dots with just THREE pen strokes!

Rotated and reflected shapes still count.
Using just three straight lines. **You should find 16 in total.**

Solution found?

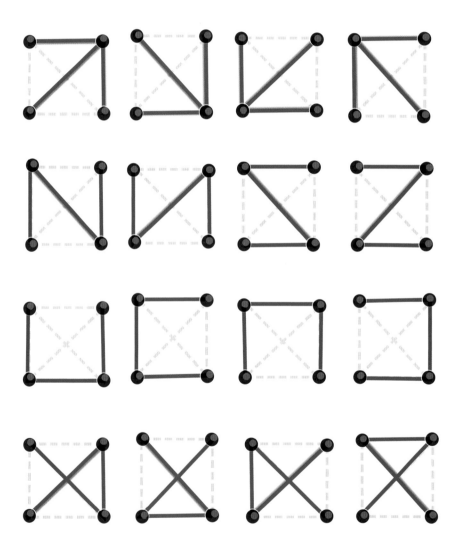

Well done if you've got this far!

Creating puzzles from the SAME shape pieces!

Using the square as our base, we found we could make another three geometry shapes. It's a bit of a maths puzzle how this works and a memory challenge.

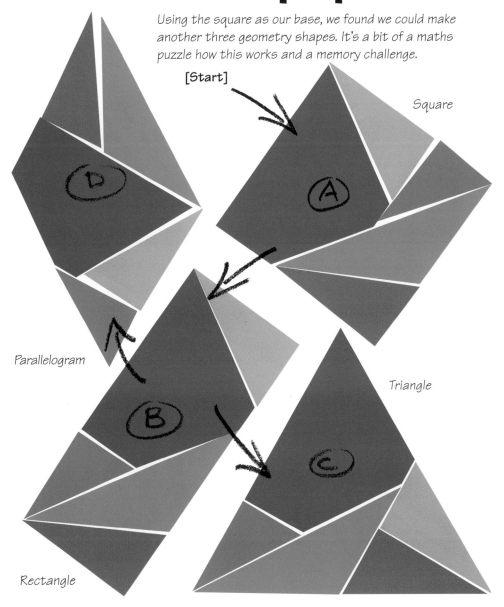

[Start]

Square

Parallelogram

Triangle

Rectangle

Photocopy (overleaf) then try to make ALL these FOUR shapes from memory.

Cut-out the pieces...

Note: be very
careful cutting
out the pieces!

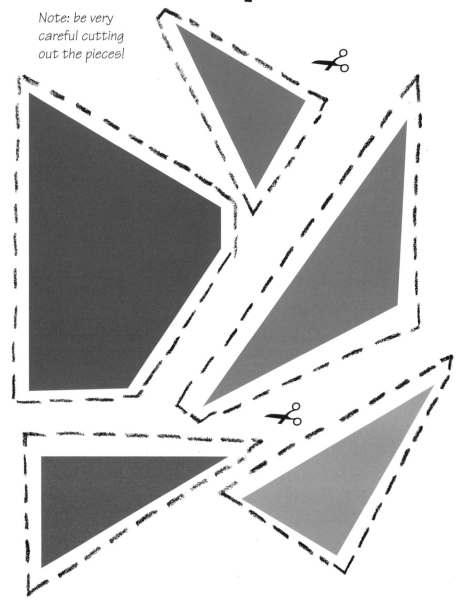

All the pieces are single-sided (printed one side only). First copy this
page (photocopier), glue and mount on stiff card, Then cut-out colours
(not grey lines). Then finally try out all four shapes you saw earlier!

■ ■ NAPOLEON'S THEOREM

Napoleon Bonaparte was famous for being a conqueror, but he was also an accomplished mathematician as he proves here.

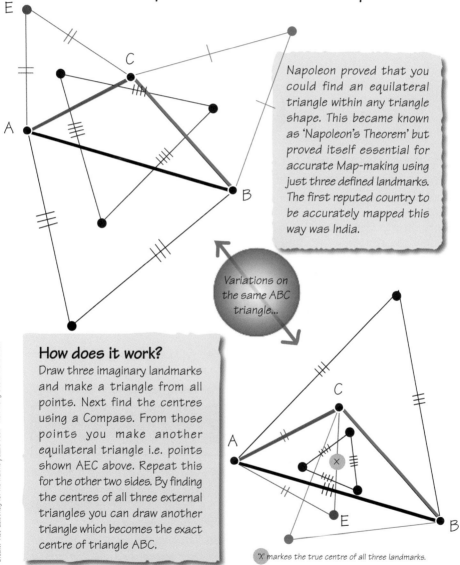

Napoleon proved that you could find an equilateral triangle within any triangle shape. This became known as 'Napoleon's Theorem' but proved itself essential for accurate Map-making using just three defined landmarks. The first reputed country to be accurately mapped this way was India.

Variations on the same ABC triangle...

How does it work?

Draw three imaginary landmarks and make a triangle from all points. Next find the centres using a Compass. From those points you make another equilateral triangle i.e. points shown AEC above. Repeat this for the other two sides. By finding the centres of all three external triangles you can draw another triangle which becomes the exact centre of triangle ABC.

'X' markes the true centre of all three landmarks.

Now all you need is a Compass to puzzle it out yourself!

More puzzling mapping techniques prove useful overleaf...

Credit: Rob Eastway for his Maths puzzle book "How Many Socks Make a Pair"

NAPOLEON △ PUZZLE MAP

English (OS) Ordnance Survey maps are based on landmarks of three points (called Beacons) to make a 'triangulation' or points. By adding the 'angles' between the points and using simple trigonometry you can find the 'centre' and the distance from each points is rather easy!

All by using just a magnetic compass (and a pair of legs)!

34°

Triangulation starts with the measured length of just one side, (which would have been walked (paced) between two points. The other two points and angles are used to calculate the distance.

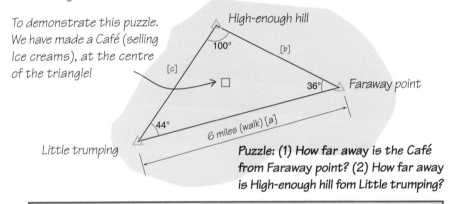

To demonstrate this puzzle. We have made a Café (selling Ice creams), at the centre of the triangle!

High-enough hill

100°

[b]

[c]

□

36° △ Faraway point

44°

6 miles (walk) [a]

Little trumping

Puzzle: (1) How far away is the Café from Faraway point? (2) How far away is High-enough hill fom Little trumping?

Easy solution: Test Napoleon's geometry skills with a ruler pencil, compass, protractor to find the hidden Café and the distance to all far points.

'Trignometry' is also widely used by Geographers and map makers when measuring the angles between the two compass points. In the hills and few mountains of England it is not uncommon to find a 'Beacon' - a 'TRIG PILLAR' on the top (usually with magnificent views). They have since become popular meeting points for walkers and hikers following the Ordnance Survey maps. 'TRIG points' are clearly indicated on maps by △ blue triangles with a dot inside, (see if you can find one on a map)....
http://www.ordnancesurvey.co.uk/blog/2012/05/trig-pillars-we-salute-you/

A mathematical solution using easy TRIG functions can be found below:
http://www.mathsisfun.com/algebra/trig-cosine-law.html

Puzzling Greek oddity

A Maths 'problem' or a Greek paradox?

Credit : based on a problem explored in Kjartan Poskitt's Murderous Maths book. www.murderousmaths.co.uk

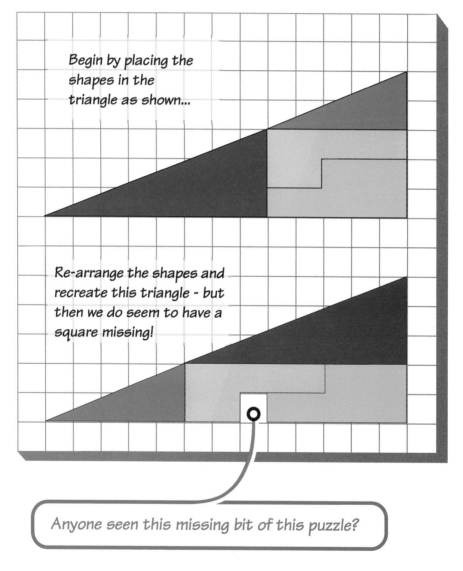

Begin by placing the shapes in the triangle as shown...

Re-arrange the shapes and recreate this triangle - but then we do seem to have a square missing!

Anyone seen this missing bit of this puzzle?

Photocopy and cut-out the page overleaf carefully and re-arrange the pieces like this. See if you can solve the puzzle (or your parents)!

ODD!

How can it be that simply re-arranging the shapes we find a square missing?

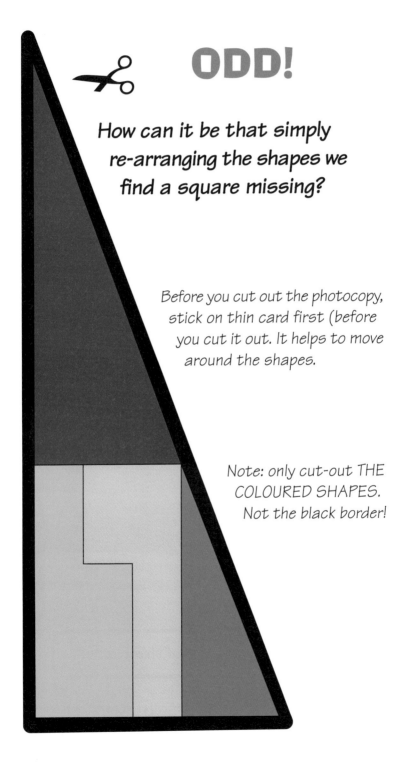

Before you cut out the photocopy, stick on thin card first (before you cut it out. It helps to move around the shapes.

Note: only cut-out THE COLOURED SHAPES. Not the black border!

The case of the missing Pencil...

Have you ever wondered where your missing pencils go? *The answer could be puzzling...*

* Count the Pencils FIRST and then cut out the dotted lines. Next swop TOP parts A and B and recount. Find out where it's missing!

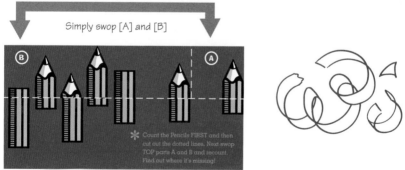

Simply swop [A] and [B]

* Count the Pencils FIRST and then cut out the dotted lines. Next swop TOP parts A and B and recount. Find out where it's missing!

Can you find THAT missing pencil?

Cut out the puzzle overleaf...

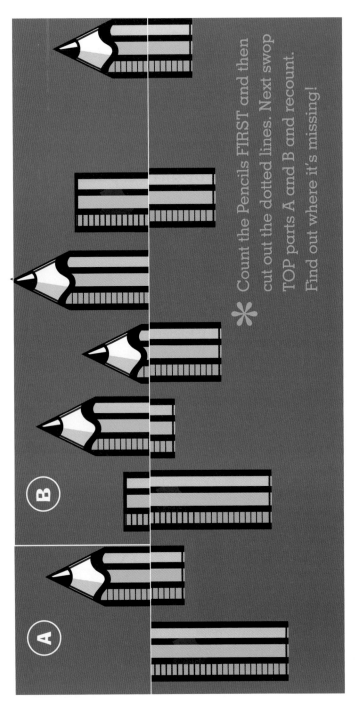

✱ Count the Pencils FIRST and then cut out the dotted lines. Next swop TOP parts A and B and recount. Find out where it's missing!

Photocopy first to avoid spoiling your book...

Cut out along the WHITE lines only!

How **LONG** is the Ladder?

This puzzle is located in the NEW 'Crossrail' London underground train tunnel. We now need to know how exactly how LONG that Ladder is and it's angle. Should be easy!

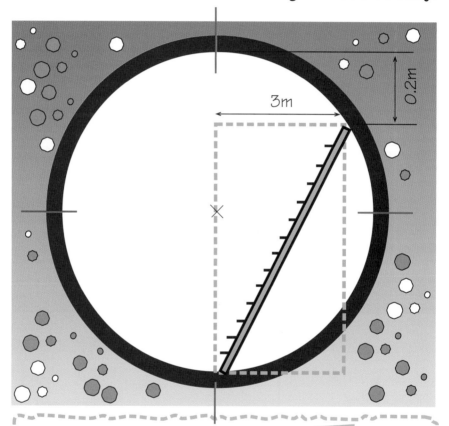

Based on an idea by Chris Maslanka (printed in The Guardian 2009).

We do know that the tunnel has a diameter of 6.2m. This puzzle presumes that the Ladder rests on the vertical centre of the tunnel [1] and the Ladder rests against the wall [2] (inclined) 6m HIGH (because 6.2 - 0.2m) it should not be too difficult. You can use Pythagoras to work out the length of the Ladder!

You can thank Pythagoras for finding this out. We think it's a bit of a puzzle and you need to be a bit clever to find the angle.

Turn the page only afer you have thought about it.

An easy puzzle when you KNOW HOW!

Pythagoras had already solved the puzzling length of that Ladder a long time ago, here's how he did it..

The diagonal of the square (above left) is the square root of two, which is **1.414213562**.... We can do this by using $a^2+b^2=c^2$ ('a' squared plus 'b' squared equals 'c' squared which is the hypotenuse). We use the same formula...

Ladder length: **6.708m*** = square root of (3x3)+(6x6)

(rounded to three decimal places)

Now we can discover the angle of the Ladder...

You could draw the geometry (to scale) and then use your protractor but it's far more accurate to use the **COSINE** rule...

Cosine rule: $c^2 = a^2 + b^2 - 2ab\, cosC$

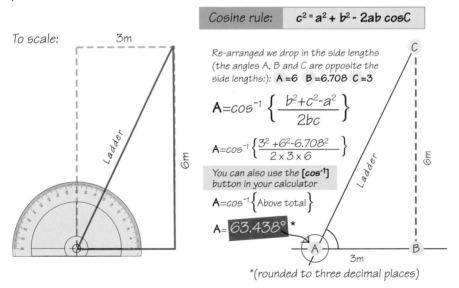

Re-arranged we drop in the side lengths (the angles A, B and C are opposite the side lengths:): **A =6   B =6.708   C =3**

$$A = cos^{-1} \left\{ \frac{b^2+c^2-a^2}{2bc} \right\}$$

$$A = cos^{-1} \left\{ \frac{3^2 +6^2-6.708^2}{2 \times 3 \times 6} \right\}$$

You can also use the **[cos⁻¹]** button in your calculator

$$A = cos^{-1} \left\{ \text{Above total} \right\}$$

A = **63.438°** *

(rounded to three decimal places)

To scale: 3m

Ladder   6m

This is very handy when you next loose your ladder in a London tunnel and forget entirely how long that ladder was! You will be able to solve other puzzles like this using the COSINE rule... You can find more of that here: http://www.mathsisfun.com/algebra/trig-cosine-law.html

Online COSINE Calculator: **Wow!** that was easy-peasy!

http://www.calculatorsoup.com/calculators/geometry-plane/triangle-law-of-cosines.php

The Roman Palindrome (found 1AD)

This LATIN puzzle was written on a wall in the doomed city of Pompei recently excavated and restored! Is it a secret code or just a bit of early graffiti?

You'll need your school Latin dictionary to read it! And a very clever word play on a square. Reading left-to-right, up and down the same words repeat. It seems to say **'THE SOWER AREPO HOLDS THE WHEEL WITH EFFORT'** but it may have meant something 'else' at the time to those early Italians. Another clever (and more recent) play on words is; **'A MAN, A PLAN, A CANAL, PANAMA',** which reads backwards as well as forwards! Not that easy to do as a 'Crossword' puzzle...

Palindromes are usually unique, but what if this is also a hidden Anagram...

Awesome (or awful) Anagrams...

Anagrams are wordplays on letters. Anagrams involve putting letters in a different order. Clever anagrams usually mean something (the Roman Palindrome on the previous page may be also an anagram; it may give the original word - a new meaning.

Example: Type one
'Mate'
Atem - (does not make sense)

Example: Type two
'Mate'
Meat - makes sense

Example: Type three
'Mate'
Team - relates to the original word

A few really clever anagrams:

The eyes	>> They see
Astronomer	>> Moon starer
Schoolmaster	>> The Classroom
Slot machines	>> Cash lost in 'em
One plus twelve	>> Two plus eleven
The National Geographic Society	
>> a great lion photo is eye-catching	
Election results	>> lies - let's recount
Halley's comet	>> Shall yet come
Willam Shakespeare	>> I am a weakish speller

However, some anagrams are based on famous sayings:

'That's one small step for a man and one giant step for mankind'.
(Neil Armstrong) >> A thin man ran... makes a large stride... left planet... pins flag on Moon... on to Mars*

Or People...
Oliver Cromwell　　>> More evil WC roll
Queen Victoria　　>> I acquire one TV
Clint Eastwood　　>> Old west action

Multiple anagrams:
Thomas Stearns Eliot
>> To the arts I am a lesson... or loathsome train sets

Well the puzzle may seem rather obvious now. All you have to do is find your favourite author and make a sentence or meaningful word from the same letters.

How hard can that be?

You can see some of these and more from two sources: Horrible Histories series by Terry Deary's 'Wicked Words' book and 'Fun with words' website: http://www.fun-with-words.com

How high is that BIG CASTLE exactly?

The Romans found out 'heights' purely on eye-level observation and a stick...

FIRST: You need to mark your stick in inches...

SECOND: Pace out 11 yards from the building, and place your stick on that mark.

THIRD: Pace one further yard and place a marker stone [a] in line with the inch stick.

?

[a]

1 pace   11 paces

An INCH: Described as a 12th of an inch 'UNICA'.
A FOOT: meant originally the length of a human foot.
A PACE: means a 'passus' which is two natural foot-steps
A YARD: a Roman 'STICK' (equals a YARD), 36 inches.

FOURTH: From the ground level, align base mark [a] to the to the top of building via the stick measured in inches..

So you want to know how HIGH the Castle is?

(1) Start by walking 11 paces away from the castle. (11 paces is left foot and right foot = 1 pace). (2) Place your inch stick on the ground and stand upright and another marker one pace away. (3) At ground level one Romain soldier looks from point [a] to the top of the building via the inch stick (4) You can measure the height by reading from the inch stick how high the building is.

Simple Maths: With 1 pace + 1 1 paces equals 12 units (the actual measurements feet, inches or metres are unimportant). The stick is marked in inches and twelve inches equal one foot. The soldier on point [a] can read off the scale of the stick and find how many inches above the ground the height actually is. **One inch on the stick = 1 foot height.**

Roman measurements: http://www.hemyockcastle.co.uk/measure.htm

Latin MAD Maths...

As you're so good at the previous puzzle we thought you would like to show off your Math skills with these four 'easy' Roman sums...

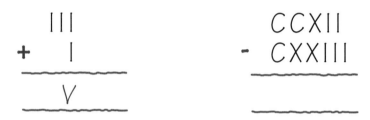

Answers should be written in roman numerical figure
I, II, III, IV, V, VI, VII, VIII, IX, X, XII and XII
You might find the Roman CALCULATOR* (at bottom of page), useful.

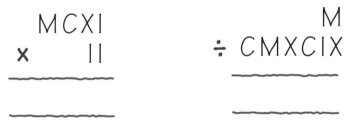

No rush - take your time!

You might also like to find the height of your house (or School) using a piece of string...

FIRST: *Use a yard long ruler (36 inches) as the stick.* **SECOND:** *holding the stick walk the distance (11 paces) from a nearby building (or Tree if you prefer)* **THIRD:** *Walk one pace more (two strides) and place a rock on the ground as marker [a]. Tie some string to the rock and hold to one eye - whilst holding the stick for balance upon the ground and looking at the top of the tree (or building) read off the height on your ruler.* **One inch=one foot in height.**

*http://romannumerals.info/roman-numerals-1-1000/

Heron *of* Alexandria
Discovers the AREA of a triangle!

Also nicknamed 'hero', he thought up this rather nifty equation for you students in about 150 B.C.E

$$A = \sqrt{s \times (s-a) \times (s-b) \times (s-c)}$$

OK it looks hard? - but NOT if you know what the letters stand for, (all parts of a typical triangle below):

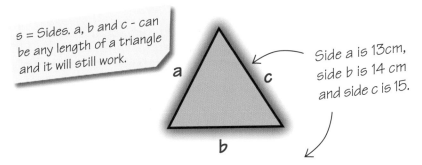

s = Sides. a, b and c - can be any length of a triangle and it will still work.

Side a is 13cm, side b is 14 cm and side c is 15.

i.e. above example we add side **a** (13), to side **b** (14) to side **c** (15) totals 42, and then halve it; is 21. This is the definition of **s**. Using the equation (at top of page) this becomes:

$$A = \sqrt{21 \times (21\text{-}13) \times (21\text{-}14) \times (21\text{-}15)}$$

$$A = \sqrt{7056}$$

$$A = 84 \quad (\text{or } 84\text{cm}^2 \text{ to be exact})$$

PUZZLE SOLVED!

Isn't that FANTASTIC? I think you had better work it out yourself - to see if it is still true! Amazing or what, eh? Try the online calculator below!

https://www.mathsisfun.com/geometry/herons-formula.html

Pythagoras *of* Samos

Discovers the SIZE of a triangle in 3D!

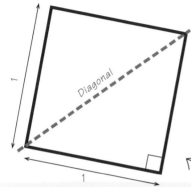

1

Diagonal

1

ONE:

Pythagoras had worked out the exact ratio of a diagonal line in 3D in 500 B.C.E. This was well before Heron found the AREA of a triangle (seen on the previous page).

The diagonal above is the square root of two, which is 1.414213562.... We can do this by using $a^2+b^2=c^2$ (a squared plus b squared equals c squared which is called the hypotenuse*).

c^2 a^2 b^2

* Greek term describing the longest side in a right angled triangle.

TWO:

Now you can easily work out the diagonal of a 3D Cube like this...

Pythagoras had created an imaginary triangle in the cube right (in red) he then created another formula for the cube diagonal...

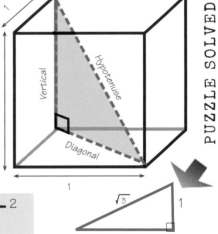

Vertical   *Hypotenuse*   *Diagonal*

1

1

1

PUZZLE SOLVED!

$$1^2 + \sqrt{2}^2 = \sqrt{3}^2$$

$\sqrt{3}$ 1 $\sqrt{2}$

...and if you work THAT out - the square root of three is:

1.732050808...

Pythagoras was the first Maths teacher and pioneer of his time.

http://www.mathopenref.com/pythagoras.html

Take 24 counters and re-arrange them into three groups

If we take 24 counters and place half - on the RED hoop, a third - on the GREEN hoop and a quarter - on the BLUE hoop...

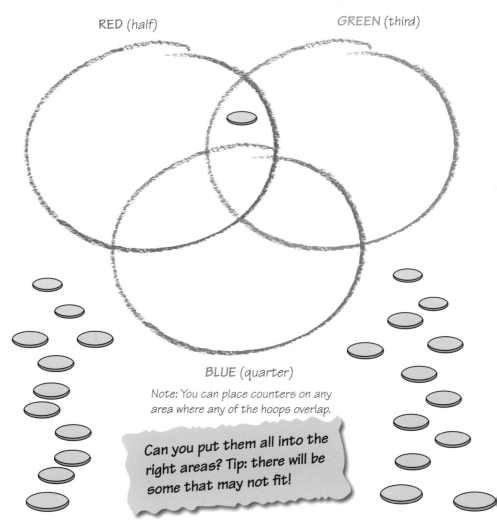

RED (half)

GREEN (third)

BLUE (quarter)

Note: You can place counters on any area where any of the hoops overlap.

Can you put them all into the right areas? Tip: there will be some that may not fit!

Although the answer is overleaf, think about how you would work this out...

How did you do?

Answer: *should be something like this...*

RED = has half of all counters   GREEN = has a third of all counters

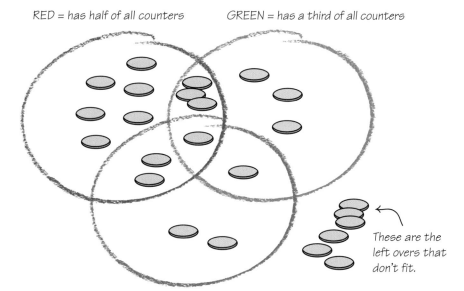

These are the left overs that don't fit.

BLUE = has a quarter of all counters

(R) Red=1/2, (G) Green=1/3, (B) Blue=1/4

Easiest way: *Find all combinations or RGB in a list - including the fractions.*

ADD

* **RGB**	= 1/2x1/3x1/4	= 24x1/24	=1
RG	=1/2x1/3	= 24x1/6	=4
RB	=1/2x1/4	= 24x1/8	=3
GB	=1/3x1/4	= 24x1/12	=2
R	=1/2	= 24x1/2	=12
G	=1/2	= 24x1/3	=8
B	=1/4	= 24x1/4	=6

(always work from most letters to least).

Tips: You have to start by dividing all the counters into the half, third and quarter sections (*as above).

SUBTRACT

* (4-1)	=3 into RG
(3-1)	=2 into RB
(2-1)	=1 into GB
(12-1-3-2)	= 6 into R
(8-1-3-1)	=3 into G
(6-1-2-1)	= 2 into B

TOTAL= 1+3+2+1+6+3+2=18

(24-18)= 6 left outside

Tips: You have to take away those that are in shared areas *(as above)

> You do the Maths. This is based on SET theory using a VENN diagram
http://passyworldofmathematics.com/three-circle-venn-diagrams/

The Polygon **HEIGHT** puzzle

Only two things you need to know are that (1) all polygons have equal sides (2) one measurement of one side is one!

1

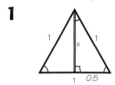

$0.5^2 + x^2 = 1^2$
$0.25 + x^2 = 1$ x = height
$x^2 = 1 - 0.25$
$x^2 = 3/4$
$x = \sqrt{(3/4)}$
$x = (\sqrt3)/2$
x = 0.86602540...

KEY FORMULAS:

Trig. formula
sinA/a = sinB/b = sinC/c

Polygon angle formula
(180(no. of sides - 2))/no. of sides

Pythagorean theorem
a^2 + b^2 = c^2

2

SQUARE

x = 1

*e.g. All side measurements
equal one. It does not matter
what units you use.*

3

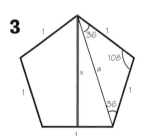

PENTAGON

$(sin36)/1 = (sin108)/a$
$a = (sin108)/(sin36)$
$a = (1+\sqrt5)/2$

$x^2 + 0.5^2 = ((1+\sqrt5)/2)^2$
$x^2 + 0.25 = (6+2\sqrt5)/4$
$x^2 = (6+2\sqrt5)/4 - 1/4$
$x^2 = (5+2\sqrt5)/4$
$x = (\sqrt{(5+2\sqrt5)})/2$
x = 1.53884177...

Proof that sin108/sin36 = (1+sqrt5)/2

sin 108 = sin 72 = sqrt((5/8)+(sqrt5)/8)
sin 36 = sin 144 = (sqrt(2(5-sqrt5)))/4

sin 108/sin 36 = sqrt((5/8)+(sqrt5)/8)/((sqrt(2(5-sqrt5)))/4)

multiply by 4 (top and bottom)
4(sqrt((5/8)+(sqrt5)/8))/(sqrt(2(5-sqrt5)))

simplify top
4(sqrt(5+sqrt5)/8)/(sqrt(2(5-sqrt5)))

bring the 4 into the square root
(sqrt(16(5+sqrt5)/8))/(sqrt(2(5-sqrt5)))

simplify top
(sqrt(2(5+sqrt5)))/(sqrt(2(5-sqrt5)))

square both sides (you MUST square root later -
remember this is squared)
(2(5+sqrt5))/(2(5-sqrt5))

simplify
(5+sqrt5)/(5-sqrt5)

rationalise denominator
(5+sqrt5)(5+sqrt5)/(5-sqrt5)(5+sqrt5)

expand
25+5sqrt5+5sqrt5+5/25-5sqrt5+5sqrt5-5

collect like terms
(30+10sqrt5)/20

simplify
(3+sqrt5)/2

multiply by 2
(6+2sqrt5)/4

square root
(1+sqrt-50)/2

factorise
((1+sqrt5)(1+sqrt5))/4

square root (undoing the square you did earlier)
(1+sqrt5)/2

4

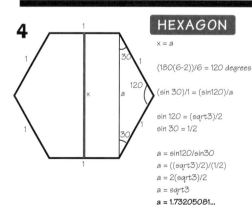

HEXAGON

x = a

(180(6-2))/6 = 120 degrees

$(sin 30)/1 = (sin120)/a$

sin 120 = (sqrt3)/2
sin 30 = 1/2

a = sin120/sin30
a = ((sqrt3)/2)/(1/2)
a = 2(sqrt3)/2
a = sqrt3
a = 1.73205081...

*By Joshua Searle
(more polygon solutions overleaf)*

The Polygon **HEIGHT** puzzle

5

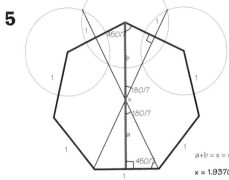

HEPTAGON

sin(180/7)/0.5 = sin(450/7)/a
2sin(180/7) = sin(450/7)/a
a2sin(180/7) = sin(450/7)
a = sin(450/7)/(2sin(180/7))

(sin(450/7)/2sin(180/7))^2+0.5^2 = b^2

b = sqrt((sin(450/7)/(2sin(180/7)))^2+(1/2)^2)

a+b = x = sin(450/7)/(2sin(180/7))+sqrt((sin(450/7)/(2sin(180/7)))^2+(1/2)^2)

x = 1.93707686...

6

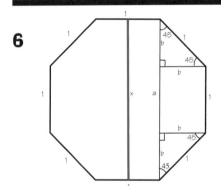

OCTAGON

x = a

2b^2 = 1^2
2b^2 = 1
b^2 = 1/2
b = sqrt(1/2)
b = 1/sqrt2
b = (sqrt2)/2

2b = sqrt2
a = 1+sqrt2

x = 2.41421356...

7

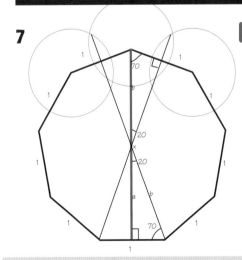

NONAGON

(180(9-2))/2 = 140 degrees

sin20/0.5 = sin70/a
2sin20 = sin70/a
a2sin20 = sin70
a = sin70/(2sin20)

(sin70/(2sin20))^2+0.5^2 = b^2
sqrt((sin70/(2sin20))^2+0.5^2) = b

a+b = sin70/(2sin20) + sqrt((sin70/(2sin20))^2+0.5^2)

x = 2.83564091...

I am sure you will find these useful to find the height of more polygonal shapes!

Ground effect aircraft

Using just a School protractor you can show how a Ground effect aircraft flies...

A well known puzzle (by generations of naughty schoolboys), is that if you place the Protractor face down on a flat surface and give it a hard knock, it will skitter across the table top before crashing to the floor...

How does it work?
The front lip of the Protractor is 'key' to directing the forward air thrust 'down'. The air sandwiched between the table top and the Protractor makes for an impressive 'Hovercraft' which makes the protractor behave as if it were weightless!

Tip: The FRONT leading lip edge must face down to fly!

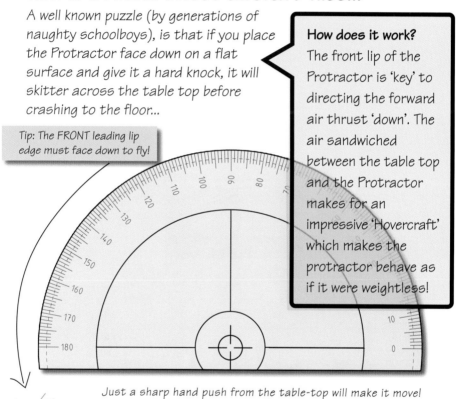

Just a sharp hand push from the table-top will make it move!

Trapped air-pressure lifts the weight so it fly's across the table...

The discovery of the **Ground effect** has led to a new generation of Airplanes designed to fly at low levels at a slow air speed. Some amazing planes have already been tested and built although don't expect to see any at your local Airport!

http://www.aviation-history.com/theory/ground_effect.htm

Not only but also;

Some puzzling aircraft that fly unlike any other aircraft (using the 'ground effect').

Normal aircraft cruise at 35,000 feet

*typical passenger aircraft cruising height .

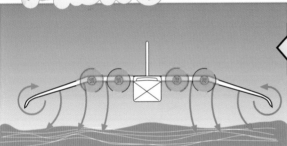

Ground effect aircraft cruise at 20 to 50 feet

The Pelican is a high-capacity cargo plane concept currently being studied by Boeing Phantom works.

Credit: Pelicam photo courtesy of Boeing Corporation (USA)

The original idea was developed in Russia during the Cold war as a heavy duty troop and tank carrier. It could lift 10,000 tons. It was never developed commercially but BOEING (of the USA) do have a picture of what they think (their own version) would look like - if they ever built one! But the massive Russion version was bulit and worked brilliantly! See Russian youtube video below.

https://www.youtube.com/watch?v=0w-Iffjhy2o

Titius-Bode law

There seems to be a mathematical formula to predict the distances of the planets. Now we know it is just a coincidence, but it was accurate enough for controversy over 'planet x'

This is how it works:

Although NO **planet** X was ever found. A recent discovery (2016) describes a new planet found deep in Space.. **See next page!**

1. Start with 0,3 then start doubling...

0, 3, 6, 12, 24, 48, 96, 192, 384, 768...

2. Add 4

4, 7, 10, 16, 28, 52, 100, 196, 388, 772...

3. Divide by 10

0.4, 0.7, 1.0, 1.6, 2.8, 5.2, 10.0, 19.6, 38.8, 77.2...

All these numbers are very close to the AU* of the planets!

*AU stands for astronomical units and 1 is the distance between the earth and the sun. You need to get a ruler and find 1cm. Imagine this as 1 AU. Now put the other distances on. You will now have the relative distances of all the planets.

Planet	Prediction	Actual	Difference %
Mercury	0.4	0.39	2.56
Venus	0.7	0.72	2.78
Earth	1.0	1.00	0.00
Mars	1.6	1.52	5.26
Ceres*	2.8	2.77	1.08
Jupiter	5.2	5.20	0.00
Saturn	10.0	9.54	4.82
Uranus	19.6	19.20	2.08
Neptune	38.8	30.06	29.08
Eris*	77.2	67.70	13.31

*Ceres is the largest asteroid in the asteroid belt, it is about 950 km in size.
*Eris is the most massive dwarf planet in the Kuiper belt, it is about 2500 km diameter.

Pluto is a bit bigger than Eris. Pluto itself is smaller than Eris in mass (not size). However if we follow the original law (the basis for our puzzle), allowing large discrepancies of 30%, the next planet (if there is one) is somewhere between 100 -200 AU... **waiting to be discovered!**

Credit for this astronomy Puzzle goes to Johann Titius in 1766.

http://www.astro.cornell.edu/academics/courses/astro201/bodes_law.htm

A new planet X has been found!

Its not a planet predicted by Titius-Bode.
But the discovery of a completly 'new planet'
found in our solar system is just as exciting
as that predicted by Titius-Bode's in 1766!

This is a very puzzling planet discovery...
It has been called a PLANET X even though nobody has seen it yet!

And your question may be what if it is not a planet but simply an
asteroid? Since Pluto (one of our farthest planets) has been replaced
by 'Eris' as a 'dwarf planet'. This new planet 'x' predicted is a radically
different Maths to that used by Titius-Bode and far more complicated.

> By the time we print this book. You may know more about it than I do.
> So here is a link to that puzzling discovery (reported by NASA).
> **https://solarsystem.nasa.gov/planets/planetx/indepth#!**

It can be a bit puzzling (even for me) so I thought I would
chart the basic differences between planets and asteroids...

1. An Asteroid or Comet is a rock or Ice that travels through space. It
may be static or usually moving (very fast). Some of these are called
Comets and sometimes we can see them at night with (ice) vapour trails.

2. Meteors and Meteorites are those 'rock' Asteroids that hit the Earth.
There are hundreds each day and they usually 'burn up' in our atmosphere
(a few do hit the ground with a thump and sometimes leave a big crater).

3. Planets are a problem. Pluto was a planet and now it a 'dwarf planet'.
And now Astronomers have accurately predicted another Planet far, far
away that is in orbit around our Sun. **Titius Bode would be pleased!**

Gridworks - draw a continuous line joining all the dots...

With a pencil start with the 3x3 dot grid and join all the dots in par 4 (that means in four lines or less). Sounds easy doesn't it?

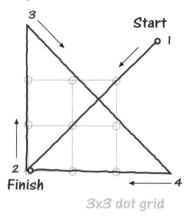

3x3 dot grid

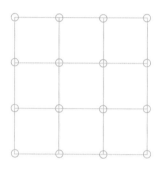

4x4 dot grid

Example: (above) join all the dots in as few lines as possible. i.e. four lines on this 3x3 grid.

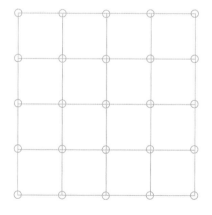

5x5 dot grid

See if you can solve all the grids before turning to our page overleaf...

Note: there is usually more than 'one way' to solve a problem and there may be a better way than the Author's solution!

Gridworks - a solution of sorts...

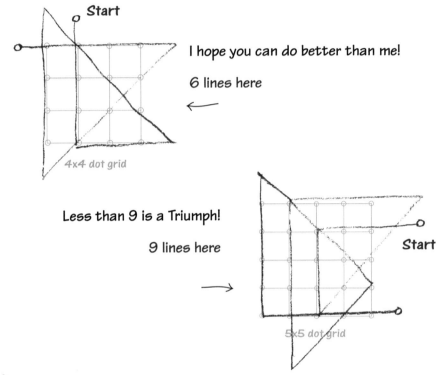

Start

I hope you can do better than me!

6 lines here

←

4x4 dot grid

Less than 9 is a Triumph!

9 lines here

→

Start

5x5 dot grid

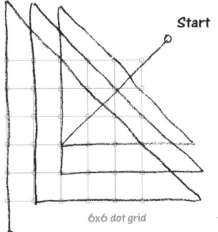

Start

6x6 dot grid

←

Well it looks as if we have done it, and we think we have. And now it is a challenge for you to perfect in less lines than we did. We do think it's possible.

10 lines here

The beauty of this game is that you can play it anywhere with just a paper and pen (or pencil)!

Face **ALL** the numbers!

A picture of the author!

I have used some fancy fonts (mixed up the typefaces) on the computer to make this puzzle picture and used every number once **(0 to 9)** to make this face, so you should find all the numbers easily. After that you can perhaps make a number face yourself using a school (or home computer drawing program) to do this. I find it helps to make a quick pencil sketch first. You also might like to find out how old this picture is by adding all the numbers up, **(see puzzle page 38)**.

DIY Letter coding machine

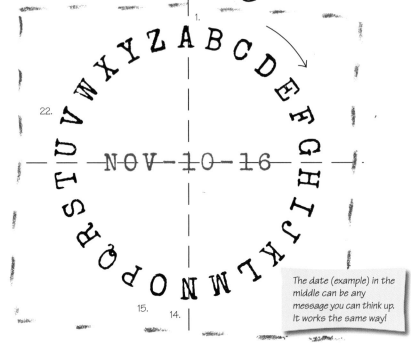

The date (example) in the middle can be any message you can think up. It works the same way!

Our own 'made-up' Letter puzzling machine

You can puzzle it out easily by making a letter wheel (like this above) and finding the opposite letter for each letter in a name. **You do need a ruler to see how it's done.** This is not random but difficult enough when you add in our (-1) rule that the direct letter is 'opposite' i.e. the opposite letter to **A** is **N**- so using our (-1) rule we get the letter **M**.

So if we start with a date **NOV-10-16** (we made up the date).

Then each letter of 'NOV' in the wheel becomes 14, 15, and 22.

After encoding 'NOV'becomes this:

Letters: ZAH

Letters are individual whilst numbers can be a problem to solve. But we could make that a 'rule' that any number has a 'Z' prefix. (Any number from 1 to 26 would have a 'Z' prefix or it could be spelt - like 'ten'). Photocopy or make your secret puzzle message!

Letters + figures: ZAH-ZV-ZB

Note: When 'un-encoding' you need to add (+1) to get back to the original (and opposite) Letter. Don't forget your recipient also needs a copy of the Letter-wheel (above) to decode your messges1

WE HAVE MADE A SECRET MESSAGE FOR YOU ON THE LAST PAGE OF THIS BOOK!

Calculator puzzle (no calculator required)

Impress your teacher with this puzzle. *You can add up all the numbers in less than 10 seconds. It's a maths puzzle so it is a BIG secret and you can then challenge your friends that you can* **'add them all up'** *before they can type it all into a pocket calculator...*

If your a good at Maths then it will take less than a minute. But what IF I were to tell you that you only had just 10 seconds! **Now that's a challenge!** *If you would like to know how to do it, see overleaf and discover how we could find out the answer without a calculator.*

How it works! (OK calculator required)

This will only work with number sets you have already calculated, so you will need a calculator to work out the totals of each set of numbers. The puzzle is making it look like they are all totally random numbers. We show you how to do it below:

```
3 6 2 4        2 6 6 5
7 3 4 2        3 2 3 9
6 2 7 5        4 5 4 2
3 6 9 6        3 6 3 6
2 7 5 7        9 5 5 2
```

[Set one]   [Set two]

Our four digit numbers are ready to go. By adding them up we can find the total - and even BEFORE we start adding them all up, we already know the answer!

That's because (secretly) we already knew each column adds up to total of exactly 19998 (that is before our mystery number is inserted on line four).

```
2 4 6 6        2 4 6 2
5 2 3 3        8 7 3 4
2 7 5 7        5 3 4 5
3 6 3 9        2 4 9 5
9 5 4 2        3 4 5 7
```

[Set three]   [Set four]

Creating your own number sets is easy, as long as you follow the rule. (1) The first three lines are added up to make a five digit number. (2) We then subtract the first three lines to get the last line number **(subtracting from 19998)**. (3) The **'made-up'** mystery number (we use) is on line 5. (4) We then ask all challengers to add up all the numbers (but we are just looking at line 4).

You can even SWOP number COLUMNS from each set - and still work out the total in less than ten seconds. **Amazing!**

We secretly prefix that number (on the fourth line) with a two and subtract the number two from the last digit in that line. So if you add up all the numbers of [Set one] you will find 3696 Just add 2 before 3696 and substract 2 = **23694.**

Simple! Now can you can create **your own sets** of numbers in the same way!

Acknowledgements

We acknowledge all contributors to this puzzle book as best we can. Our inspiration comes from many places, including many historic puzzles that we find interesting enough to include here. *All illustrations are by the book author.*

Lets start with the front cover. The younger age is eight and four and a half months, and you may be up to seventeen and nine months (if you worked it out correctly) but you can be any age if you understand the puzzles.

Page 9   Credit: Roald Dahl's book *'Charlie and the Chocolate factory'*.

Page 11   Credit: London's Toy Museum *was our inspiration.*

Page 13   Credit: Old Newspaper clipping. *(Author sadly lost).*

Page 15   Credit: Henry Ernest Dudeney *'Amusements in Mathematics'* 1917

Page 17   Credit: Dave Tuller for his graphic work.

Page 19   Credit: Ancient Egyption geometry *(5,000 years BC)*

Page 29   Credit: Joshua Searle + Fridge magnet demo

Page 21   Credit: Lewis Carroll *(1898)*

Page 23   Credit: Chinese mind challenge game *(we think).*

Page 25   Credit: Paul Simons, for his 'Weather' article in *The Times (2014)*

Page 27   Credit: Leonardo da Vinci *(1519)*

Page 29   Credit: Lewis Carroll *(1898)*

Page 31   Credit: Henry Ernest Dudney *'Amusements in Mathematics'* 1917

Page 33   Credit: Natural Chemisty *(H2O)*

Page 35   Credit: Albrecht Durer *(German Artist 15th century)*

Page 37   Credit: Ian Livingstone and Jamie Thompson book. *(by permission)* "How Big is your Brain?" published by 'The Independent'

Page 39   Credit: Victorian puzzle, original author unknown

Page 40   Credit: Kjartan Poskitt. *(Reproduced with his permission).*

Page 41   Credit: Eratosthenes *(205 BC) Librarian.*

Page 43   Credit: BBC 'Focus' Science magazine puzzle *(2010)*

Page 44   Credit: Science puzzle *(unknown author)*

Page 45   Credit: Nigel Corrigen *(by persmisson)* at the *dozenalsociety.org.uk*

Page 47   Credit: Arthur Guttenberg *(1883) the first Letterpress*

Page 49   Credit: Joshua Searle

Page 55   Credit: Burkard Polster's book on Anagrams *'Eye twisters'*.

Page 57   Credit: Lewis Carroll *(1917)*, illustration by Sam Loyd *(1898)*.

Page 59   Credit: Napoleon Bonaparte *(1821)* - *in part (we think)*

Page 61   Credit: Henry Ernest Dudeney *'Amusements in Mathematics' 1917*

Page 63   Credit: Tony Crilly's book *(professor Maths at Middlesex University)*
           '50 mathematical ideas you really need to know'

Page 65   Credit: Eleanor Searle at the *Holy Ghost School (homework question)*

Page 67   Credit: Pythagoras of Samos* *(569 to 475 BC)*

Page 69   Credit: Blaise Pascal *(1653)*

Page 71   Credit: Pythagoras of Samos* *(569 to 475 BC)*

Page 73   Credit: Eleanor's primary School Number puzzle. *(author unknown)*

Page 77   Credit: Philip Searle

Page 75   Credit: Henry Ernest Dudeney *'Amusements in Mathematics' 1917*

Page 78   Credit: Daily Mail *'Mindbenders'* inspired puzzle

Page 79   Credit: Edward H Adelson *(professor at MIT*)*

Page 81   Credit: Henry Earnst Dudney *'Amusements in Mathematics' 1917*

Page 83   Credit: *Author unknown.*

Page 85   Credit: Joshua Searle

Page 87   Credit: Leonard Euler *(1736)*.

Page 89   Credit: Joshua Searle

Page 90   Credit: *Ursuline high School* Library. *(Source: Cambridge University)*

Page 91   Credit: Lewis Carroll *(we think)*.

Page 93   Credit: Philip Searle

Page 94   Credit: Joshua Searle

Page 95   Credit: Philip Searle

Page 97   Credit: Joshua Searle

Page 99   Credit: Dr. Percy Spenser *(for discovering Microwaves in 1946)*

Page 101   Credit: Joshua Searle

Page 103   Credit: Philip Searle

Page 105   Credit: Joshua Searle

Page 107   Credit: Humphrey Davy *(1829) of the Royal Institute in London*

Page 109   Credit: Louis XIV, king of France *(style of French Carriage clock)*

Page 111   Credit: Erasto Mpemba *(Science pupil)*

Page 113   Credit: Joshua Searle

Page 115   Credit: Adam Hart-Davis *(from a talk in the RI* lecture theatre hall)*

Page 117   Credit: Philip Searle

Page 119   Credit: Kjartan Poskitt. *(by persmisson) from Murderous Maths books**

Page 121   Credit: Joshua Searle

Page 123   Credit: Joshua Searle

Page 125   Credit: Joshua Searle

Page 127   Credit: Rob Eastaway for his Maths puzzle book; *(by permission)*
            "How Many Socks Make a Pair"

Page 129   Credit: Kjartan Poskitt. *(Murderous Maths books)**

Page 131   Credit: Philip Searle

Page 131   Credit: Philip Searle

Page 133   Credit: Chris Maslanka *(by permission) based on his original idea.*

Page 141   Credit: Joshua Searle
            "Based on A level exam revision question"

Page 137   Credit: A Roman soldier *(anonymous)*

Page 139   Credit: Heron of Alexandria *(10-70 AD)*

Page 140   Credit: Pythagoras of Samos* *(569 to 475 BC)*

Page 142   Credit: Joshua Searle

Page 143   Credit: Joshua Searle

Page 145   Credit: Philip Searle

Page 146   Credit: Schoolboys *(all of them)*

Page 147   Credit: Johann Titius *(1766)*

Page 149   Credit: Joshua Searle

Page 151   Credit: Philip Searle

Page 153   Credit: Blaise Pascall *(invented a basic mechanical calculator in 1642).*
The original author of this pocket calculator puzzle is unknown.

*Reproduced by kind permission of Kjartan Poskitt (www.murderousmaths.co.uk)
Additional puzzle credits shown as a historical link. Many of the puzzle pages have
additional web links (all checked as being suitable for children to problem solve).
Also thanks to inspiration from the *London Toy Museum- www.vam.ac.uk/moc (page 11),
*RI-Royal Institute in London (www.rigb.org) - (page 107), plus *MIT - Massachusetts
Institute of Technology (page 79) are among other inspirations for this book credits.

The Bread rolls puzzle

Can you remember how to change these Bread rolls to appear to point DOWNWARDS by moving just two pieces?.

The answer was found on page 21.

Notice how the bread rolls are almost hexagonal in the shape here. This often happens when a circle is distorted by another circle of the same strength and density. It reminds me of a Bee-hive honeycomb - but HOW would a humble bumble bee know *anything* about hexagons?

Maybe they can do Maths!

http://www.benefits-of-honey.com/honeycomb-pattern.html

Finally our last secret message:

—

FMWQ-NKQ-KASGDF

BXNOQ-XAXXK-EFUOW-UZ-XUP

BGF-UZ-RDUPSQ-FQZ-TADGDE

QZVAK-EFAB

(See page 151 to find the secret message)

Enjoyed this Book?

Look on www.tarquingroup.com for more

Titles of interest include include:

A Puzzle a Day, Mathematical Snacks, Geometry Snacks, A Week's Problem, Who Tells the Truth, plus puzzles for all ages.

Eight puzzles in this book have been made into wooden puzzles for children to try to assemble* (and 'colour-in') at school or at home. You can (of course) make your own puzzles based on this book using just paper and cardboard, once you have inspiration! If you need templates you might find some here:

www.puzz-d.co.uk